Basics of Metamorphic Petrology

R.M. Manickavasagam

ATLANTIC
PUBLISHERS & DISTRIBUTORS (P) LTD

ATLANTIC
PUBLISHERS & DISTRIBUTORS (P) LTD
7/22, Ansari Road, Darya Ganj, New Delhi
Tel.: +91-11-4077 5252, 2327 3880
E-mail: orders@atlanticbooks.com
Web: www.atlanticbooks.com

Published by Atlantic Publishers & Distributors (P) Ltd 2024

Printed & bound in India by Atlantic Print Services

Preamble

This book is written for under-graduate and post-graduate students of Geology. After teaching metamorphic petrology for about two decades in Earth Science department, Indian Institute of Technology, Roorkee, Uttrakhand state, I realised that the students should have knowledge in the basic concepts of any subject, so that they can improve further up in their higher studies without much difficulty. With this view, the book aims to make the students understand easily the subject even without the assistance of a teacher. At the time of teaching metamorphic petrology, I followed the books written by Turner (1968), Winkler (1978), Miyashiro (1972), Yardley (1990), Philpotts (1994), Spear (1993) and others, and some relevant research papers of metamorphic petrology. At the time writing the book, I came across two more books, one by John D. Winter (2012) and another by R.S. Shrama (2016) on metamorphic petrology. Thermodynamic application part is from the books of Wood and Fraser (1976) and R.S. Sharma (2016). In writing the book, the text and figures are taken from all the above-mentioned authors books, and modified or simplified wherever necessary. Some of the rock photographs are taken from the Alex Strekesen paper available in the net along with my own photographs. The book is totally for students only, and therefore, the text and figures are not totally of my own.

While teaching the subject I followed my own method to make the students understand the subject easily. To some extent, I liked the approach of Winkler, and therefore, I adopted somewhat similar approach in teaching as well as in writing this book. Latest information, I got it from Winter and Sharma books, that I incorporated in the text and figures. I have not given much emphasis on petrogenetic grids for various lithologies in this book, as it will create confusion in understanding the basic philosophy of metamorphic petrology. To give an idea about the petrogenetic grid, I have given some information on migmatite melting part in Chapter 11. Thermodynamic chapter is short and only the basics are given.

Since the subject is very much field oriented, I tried to give as many photographs as possible so that students can benefit in identifying the metamorphic rocks in the field by using this book.

I have mentioned all the aspects about this book, and if anything is left mentioning, I own the responsibility.

Dr. R.M. Manickavasagam

Abbreviations

Act : Actinolite
Ab : Albite
Alm/Alm grt : Almandine
Anl : Analcite
And : Andalusite
As : Aluminosilicate
Ath : Anthophyllite
Arg : Aragonite
Aug : Augite
Bru : Brucite
Bt : Biotite
Cal : Calcite
Chl : Chlorite
Cld : Chloritoid
Crd : Cordierite
Cpx : Clinopyroxene
Cum : Cummingtonite
Di : Diopside
Dol : Dolomite
Ep : Epidote
Fo : Fosterite
Grt : Garnet
Gln : Glaucophane
Grs : Grossular
Hbl : Hornblende
Hul : Hulandite
Hyp : Hypersthene
Ilm : Ilmenite
Jd : Jadeite
Kln : Kaolinite
KF : K-feldspar
Ky : Kyanite
Lmt : Laumontite
Lws : Lawsonite
Mt : Magnetite
Ms : Muscovite
Omp : Omphacite
Opx : Orthopyroxene
Prg : Paragonite
Per : Periclase

Pl	: Plagioclase	Sil	: Sillimanite
Pre	: Prehenite	Stp	: Stlilpnomalene
Pmp	: Pumpellyite	St	: Staurolite
Prp	: Pyrope	Tlc	: Talc
Prl	: Pyrophyllite	Ttn	: Titanite
Qtz	: Quartz	Tr	: Tremolite
Ru	: Rutile	Wo	: Wollastonite
Scp	: Scapolite	Zo	: Zoisite

Contents

Chapter 1

Metamorphism

1.1 Introduction

Metamorphism can be defined as re-crystallization of the pre-existing crustal rocks, namely the protolith or the parent rocks. Hereafter, it is referred as parent rock. It may be igneous, sedimentary or low-grade metamorphic rock. Metamorphism of the parent rocks show changes in the original mineralogy as well as in structural/textural features. This happens only when the parent rock is subjected to new set of physical, i.e. temperature and pressure, and chemical condition that is entirely different from the conditions under which the parent rock was formed. The new set of conditions is created in the earth's crust time and again by tectonic activities or movements within the earth. The tectonic activity in the earth's crust may be any one or combination of the following factors. The earth's activities may be (a) a simple burial of the basin sediments with consequent heating, (b) an igneous intrusion causing heating of the country rock, (c) large scale (regional) tectonic activity inducing subduction of crustal sediments to depth with concomitant heating and increase in pressure, and (d) heating of surface rocks by rare meteoric impact. The crustal rocks, i.e. the parent rocks that undergo metamorphism may be purely igneous (formed > 900°C) or sedimentary (formed < 200°C) or low-grade metamorphic (formed between 200 and 400°C). In general, it has been observed that metamorphism of the protolith takes place at 200° to 800°C temperature and 1 to 10 kb pressure. During the whole process, whole rock chemical composition of the parent rock is more or less retained, except for the removal or addition of volatiles like H_2O, CO_2, F, Cl, etc. The texture of the parent rock may be retained or new textures may be developed depending on the physical as well as tectonic conditions prevailed at that time. During metamorphism, the rock mineral equilibrium is disturbed and the minerals become unstable. Coexisting minerals tend to react and produce a new set of stable equilibrium mineral assemblage. The entire metamorphic process takes place in solid state without any intermediate liquid state, unlike igneous rock crystallization from liquid magma.

Chapter 2

Factors of Metamorphism

Metamorphic process is influenced by many factors that are to be understood. It includes tectonic setting, temperature-pressure (physical) condition and whole rock chemical as well as mineralogical composition of parent rock. These factors are described in short detail in the following section.

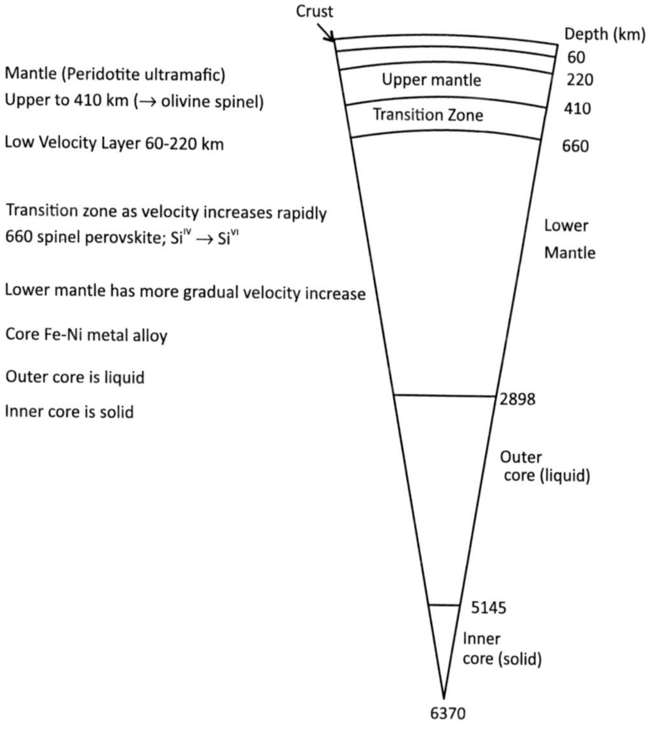

Fig. 2.1: Earth's Interior

2.1 Tectonic Settings

The outer cover of the earth is a very thin layer and is known as crust, when compared to the earth's inner part mantle and core (Fig. 2.1). Figure 2.1 depicts the cross section of the earth and the thickness of mantle and core.

The earth's crust is made up of several major and minor tectonic plates (refer to books on plate tectonic for further details). The crust is broadly divided into oceanic crust and continental crust. The oceanic crust has an average thickness of 5 km with a maximum of about 10 km. The oceanic crust is located in Pacific, Atlantic and Indian oceans. The continental crust has an average thickness of about 45 km with a maximum of about 60 km in certain part of the continents. The continental crust is located in North and South America, Africa, Europe, Asia and Australia. Oceanic plates are unstable whereas continental plates are relatively stable. The plates are either created or destructed. In the middle part of oceanic crust, the crustal plates are created, called as divergent plate margins or mid-ocean ridges (Fig. 2.1). Divergent plates are created continuously in the middle of Pacific, Atlantic and Indian oceans. The plates created at the mid-ocean ridges are destroyed at the ocean-continent, continent-continent and ocean-ocean margins known as the destructive plate margins or the so-called subduction zones. Examples of various subduction zones include the ocean-continental subduction zones of west American coast, Mediterranean region; continent-continent subduction of the Himalaya; and the ocean-ocean subduction of the Japanese and Indonesian islands.

In oceanic rift zones, effect of temperature is more prevalent with almost no effect of pressure. The temperature effect on the oceanic ridges is due to the upwelling magma from the mantle that causes metamorphism.

Plate Tectonic settings—Igneous Genesis

1. Mid-ocean Ridges
2. Intra-continental Rifts
3. Island Arcs
4. Active Continental Margins
5. Back-arc Basins
6. Ocean Island Basalts
7. Miscellaneous Intra-continental Activity Kimberlites

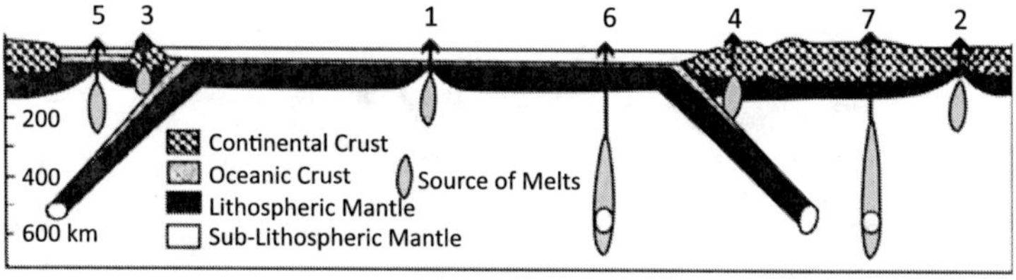

Fig. 2.2: Generalized Plate Tectonic Settings

In the subduction zones, the oceanic crust is subducted beneath the continental crust; similarly, one continent is subducted beneath another continent or an oceanic crust is subducted beneath another oceanic crust (Fig. 2.2). Various components of the subduction is presented in Fig. 2.3.

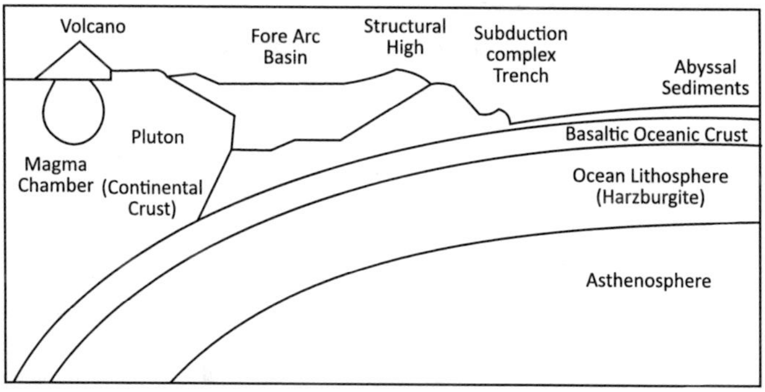

Fig. 2.3: Generalised Subduction Zone Components

The isotherms in subduction zone (Fig. 2.4) and various metamorphic facies of the subduction zones are shown in Fig. 2.5. In such zones, the crustal material is dragged downwards into the mantle. Due to its downward burial, both pressure and temperature increase with depth (Figs. 2.3 and 2.4).

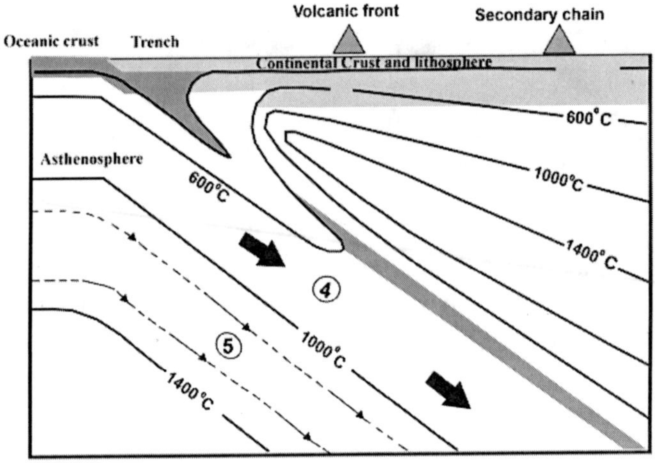

Cross section of subduction zone showing isotherms.

Fig. 2.4: Isotherms in Subduction Zone

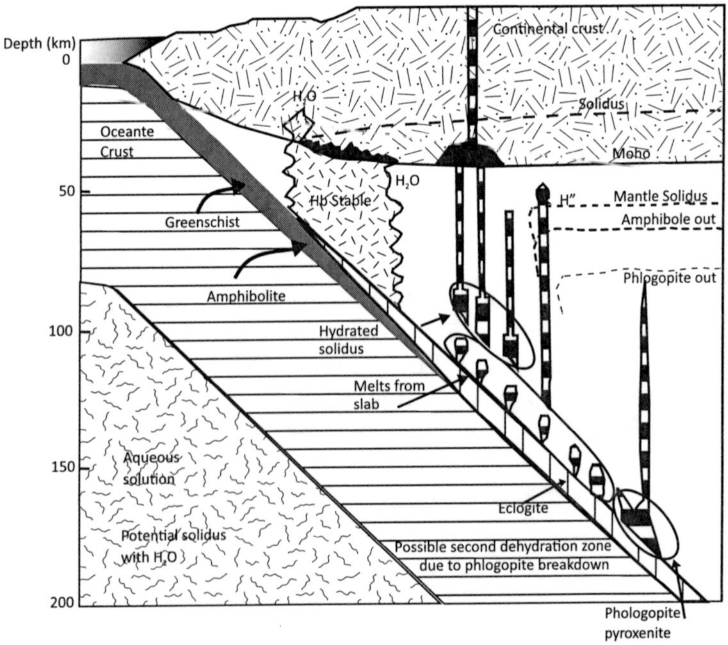

Fig. 2.5: Subduction Zone Showing Magmatism and Some of the Metamorphic Facies

The P-T increase is not uniform in all subducion zones and is dependent on the angle of subduction. Low angle subduction witness lower temperature of ~200 to ~500°C and pressure of ~1.0 to ~6.0 kb. Steeper subduction zones have higher temperature of ~400 to ~800°C and pressure of ~1.0 to ~10.0 kb. The temperature increase in shallow subduction zone is low (10 to 15°C/km) and temperature equilibration is sluggish due to the presence of silicate minerals in crustal rocks. However, in high angle subduction zones, the geothermal gradient is ~25 to 30°C/km, and sometimes, this gradient is as high as 60°C/km in certain parts of the crust due to hot spot activities. The pressure increase is due to overburden thickness and rock density. At low angle subduction zones, the thickness of the overburden is less than steep angle subduction zones. Overload pressure is low in gentler subduction zones and high in steeper subduction zones. If the density of the crustal rock undergoing subduction is 2.7 gm/cm^3, at 10 km depth the pressure is ~2.7 kb; at 20 km depth ~5.4 kb. In case, the density of the rock is ~3.3 gm/cm^3, then at 10 km depth it is ~3.3 kb and at 20 km ~6.6 kb.

2.2 Physical (P-T) Condition

Temperature (T): As mentioned earlier, when an area is affected by higher P-T condition, it undergoes metamorphism. Metamorphic activity is enhanced by presence of fluids like H_2O, CO_2, F, Cl, etc. present in the parent sediments or rock. It has been observed that the geothermal gradient in the lower to middle continental crust is lower than that of the oceanic crust (Fig. 2.6). Temperature increase in continental and oceanic crusts are relatively insufficient to produce significant metamorphism. In order to get effective metamorphic changes, some additional heat is required.

The additional heat may be derived from any one or combination of these perturbations such as (a) magmatic intrusion, (b) hotspot/plume activities, (c) radiogenic heat, (d) crustal extension, (f) mantle de-lamination, etc. which are of transient phenomenon (Fig. 2.7).

As mentioned earlier, from earth's surface towards mantle, temperature increase is ~25 to 30°C/km in continental crust and lesser in oceanic crust. In some tectonics zones, it may be as high as 60°C/km. Considering the average temperature increase in continental as well as in oceanic crust, it may or may not attain significant metamorphic changes. For example, if the temperature increase is ~25°C/km, and at 20 or 25 km depth, temperature is ~500 or ~625°C. Such temperature is sufficient to produce low to medium grade metamorphism. To produce high-grade metamorphism, some additional heat is required. The additional heat comes from (a) by hot mantle convection taking place below the lower crust, (b) hot up-welling magma moving towards the surface through crustal rocks and (c) radiogenic heat produced by crustal rocks (Fig. 2.7). Such additional heat generated by any or combination of the above-mentioned processes enhance the metamorphism.

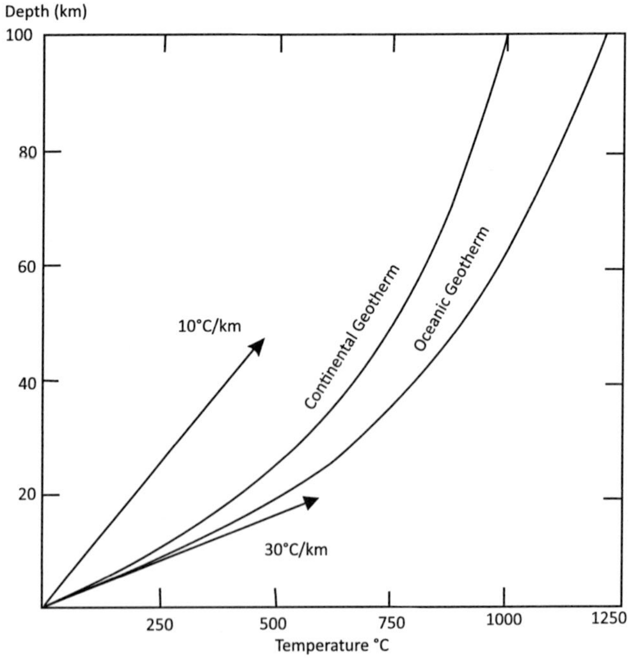

Fig. 2.6: Geothermal Gradient for Continental and Oceanic Crust

Temperature increase results in re-crystallization of the parent rock. Fine grained rocks are converted to coarse grained rocks, in a static environment without any stress, e.g. limestone to marble, siltstone or clay stone to quartzite, etc. Grain size tends to increase but beyond a certain point the grain growth is arrested. The parent rock equilibrium is disturbed due to temperature increase. The minerals of the parent rock become unstable thereby initiating chemical reaction between minerals and produce stable equilibrium mineral assemblage for the new temperature condition.

Pressure (P): Two types of pressures act during metamorphic process. They include hydrostatic pressure and directed pressure. Hydrostatic pressure is defined as the pressure exerted by the overlying column of the rock at depth. This pressure at a given depth is based on the thickness of the rock column, density of the rock and the gravitational force of the earth. The average pressure increase varies from 250 to 300 bars/km. The continental crustal rocks have an average density of ~2.7 to 2.8 cm^3/g, whereas the oceanic crustal rocks have an average density of ~3.0 cm^3/g. The low density rocks of continental crust exert pressure of ~250 bars/km. High density rocks of the oceanic crust exert pressure of ~300 bars/km. If you consider a depth of 10 km in the continental crust, the hydrostatic pressure is about 2.5 kb. At a depth 20 km, it is 5.0 kb. Similarly, in oceanic crust, at 10 km depth the pressure is about 3.0 kb and at 20 km depth is about 6.0 kb. The hydrostatic pressure is equal in all directions (both vertical and horizontal). This means that the vertical pressure exerted by the overlying column of the rock at 10 km depth is resisted by the same amount of pressure in the opposite direction by the column of the rock that is occurring below 10 km depth. In this process, the individual minerals and the pore space fluids of the overlying as well as the underlying rocks experience the same amount of hydrostatic pressure in all directions. Since the rock exerts pressure, it is known as lithostatic pressure or hydrostatic pressure or total pressure.

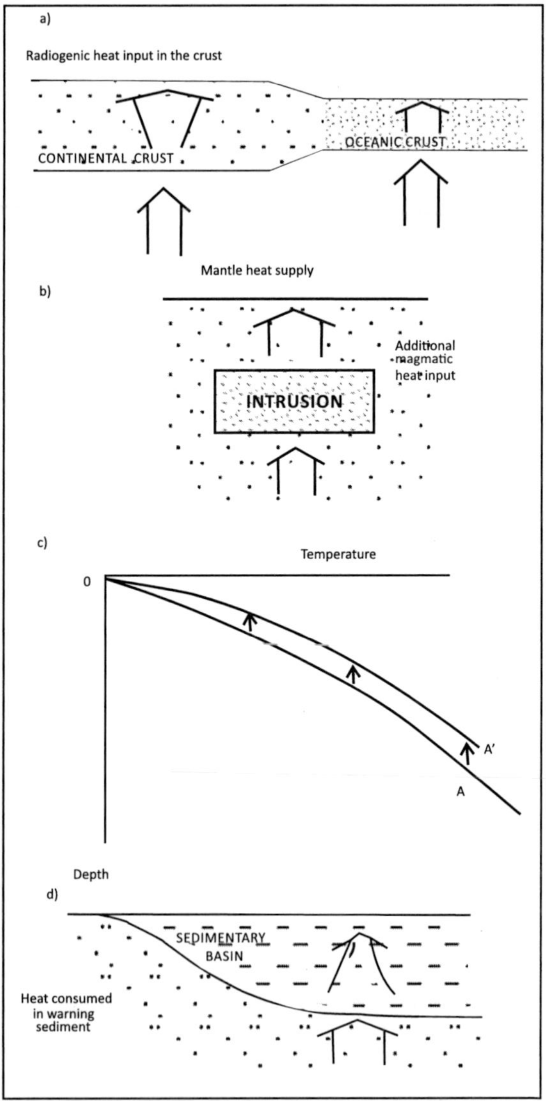

Fig. 2.7: Heat Generation from Various Sources

The second pressure is directed pressure or compressive pressure, which is observed in many orogenic or tectonic environments. When such pressure is applied to a rock column, differential pressure is developed in all three directions. In one direction, the directed pressure will be more than the other two directions. The rock is subjected to stress and gets yielded in the direction of the maximum directed pressure. In that process, the rock is deformed due to shearing stress. In orogenic environments, the effect of lithostatic and directed pressures are noticeable in the rocks affected by metamorphism. However, when the metamorphism takes place at shallow depth, as in contact metamorphism, only the effect of temperature is evident with no effect of hydrostatic and directed pressure. Lithostatic pressure at depth produces more denser minerals such as garnet, kyanite, sapphirine, omphacite, etc. depending on the temperature. If the rocks are affected by shearing stress together with overload pressure, significant structural/textural as well as mineralogical changes occur.

Fluids (f): The sedimentary rocks such as claystone, siltstone, sandstone, shale, greywacke, arkose, conglomerate, quartzo-feldspathic sediments, etc., have pore spaces that are filled by liquid mainly H_2O, sometimes with dissolved species like CO_2, Cl, CH_4, SO_4, etc., in H_2O. Weathered igneous rocks such as kaolinite (weathered granite), serpentinite (weathered basic and ultrabasic rocks like dunite, peridotite, gabbro), etc., have mainly H_2O with small amount of other volatiles in their pore spaces. When these rocks are metamorphosed, the pore space water in the initial stages at low P-T condition gets converted to water vapor, i.e., a gas. At higher P-T conditions, gas is converted into a fluid. Fluid is physically a gas having the chemical property of a liquid. In other words, it is neither a gas nor a liquid, and therefore, it is called as a fluid. In Fig. 2.8, H_2O is converted to a fluid at 374°C and 224 bar pressure that is called as critical point or second boiling of water.

Fluids play an important role during metamorphism. If H_2O is present, it enhances the metamorphic reactions and acts as a catalyst for the reactions to go for completion. If other volatiles like CO_2, Cl, CH_4, SO_4 are present more than H_2O, the reactions may not go to completion or the reactions may be arrested at certain P-T condition.

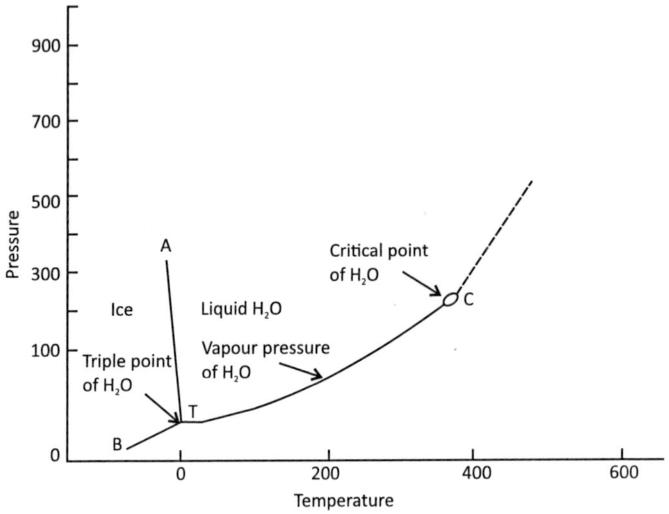

Fig. 2.8: Water at the Critical Point Shown as Open Circle—Fluid Water

2.3 Types of Change

Two major changes are observed during metamorphism. The first is the change in mineralogy, where the existing minerals are changed to new metamorphic minerals through chemical reaction in the solid state. Secondly, the existing minerals are re-crystallized producing new structures/textures without the formation of any new minerals. But on many occasions, these two changes may occur together or may take place independently based on the type of metamorphism.

One important aspect is that metamorphism takes place in solid state. Original minerals are changed to new minerals in solid state. In the entire process, the rock never undergoes to liquid state, except during partial melting at very high grade condition. In the whole process, the composition of the parent rock is almost retained, except for the removal of volatiles H_2O, CO_2, etc. Hence, metamorphism is termed as near-isochemical or isochemical.

Chapter 3

Types of Metamorphism

Several types of metamorphism are described in literature based on detailed observations made in several parts of the world for more than two centuries. They include (a) contact metamorphism, (b) cataclastic metamopshim, (c) hydro-thermal metamorphism, (d) burial metamorphism, (e) pyrometamorphism, (f) ocean-floor metamorphism, and (g) regional dynamo-thermal or regional metamorphism. Among these, the first five types of metamorphic rocks are formed on a limited scale of 1 to 10's of kilometers whereas the last two are found to occur on the regional scale of 100 to 1000s of kilometers. Ocean-floor, contact and regional metamorphic rocks are commonly found rocks in the crust.

3.1 Contact Metamorphism

When an igneous body intrudes, either from upper mantle or lower crust (~60 to 30 km depth) to near surface part of the crustal (< ~5 km depth) country rock, heat from the igneous body is dissipated into the cooler country rock causing contact metamorphism (Fig. 3.1). This type of metamorphism is commonly found close to granitic intrusions, rather than with basic intrusions. High amount of volatiles H_2O, Cl, F, etc., present in the granitic intrusion transfer heat to the country rock. Close to the intrusive contact, country rock undergoes the highest grade metamorphism (~800°C) and farther away from contact the lowest grade metamorphism (~200°C). Since the metamorphism takes place at shallow depths, the effect of pressure is negligible. Contact metamorphism produces only mineralogical change with no structural change in the country rock.

3.2 Cataclastic Metamorphism

This type of metamorphism is confined to the near vicinity of overthrusts or faults. The rocks present in the thrust zone are sheared, mechanically crushed and recrystallized. In the process, the rocks show significant structural changes with no mineralogical changes due to low heat generation. This type of metamorphism is also termed as shear-zone metamorphism or dislocation metamorphism or high stress metamorphism or a variety of dynamic metamorphism.

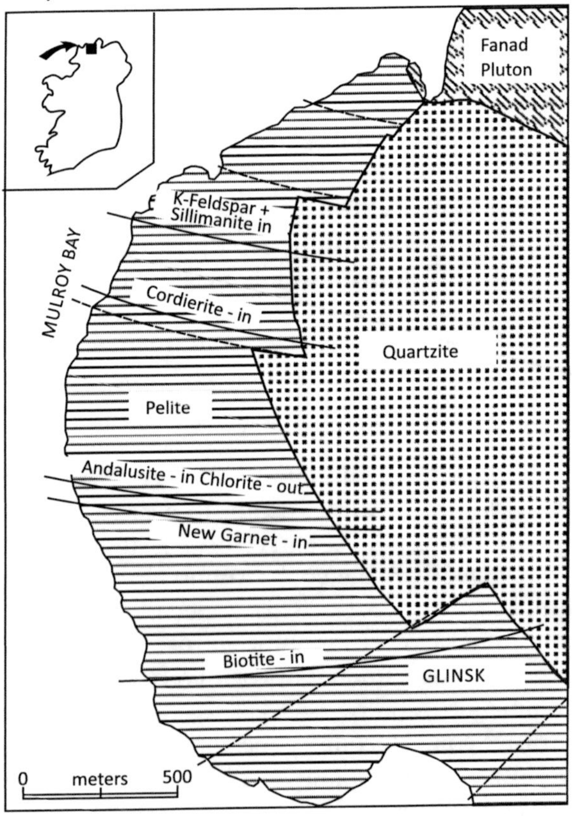

Fig. 3.1: Contact Metamorphism

3.3 Hydrothermal Metamorphism

When hot hydrothermal solution emanating from volcanic vents percolate through fractures of the crustal (mainly oceanic) rock, some mineralogical changes are noticed on a local scale in the crustal rock. Such changes are termed as hydrothermal metamorphism by Coombs (1963). The temperature is ~100° to 400°C for this kind of metamorphism. This type of metamorphism is found close to the trench region of the subduction zones, where hydrothermal metalliferrous deposits are common, e.g. Pb-Zn porphyry and Cu-Mo porphyry deposits.

3.4 Burial Metamorphism

The continental and volcanic arc sediments are accumulated in geo-synclines and buried to depth. The buried sediments are affected by low temperature, due to compaction and overloading, and show some mineralogical changes with no structural changes. It is termed as burial metamorphism by Coombs (1961). Maximum temperature attained in such metamorphism is about ~450°C. However, such metamorphism is no more valid in the present day, as the geosynclines (trench) are part of active tectonic trench zones.

3.5 Pyrometamorphsim

It is similar to contact metamorphism but found only on a limited scale of few millimeters to centimeters. It occurs close to the contact of the country rock with mafic dykes and sills of igneous origin. The country rock shows burnt appearance at the contact zone. This type of metamorphism produces rather rare high temperature minerals, commonly found in sanidinite facies (see facies chapter for details).

3.6 Ocean Floor Metamorphism

The volcanic rocks formed at mid-ocean ridges are affected by circulating hot sea water, producing low temperature minerals in such rocks. Since it is occurring at the ocean floor, it is termed as ocean floor metamorphism (Miyashiro et al., 1971) or ocean-ridge metamorphism (Spear, 1993).

3.7 Regional Metamorphism

This type of metamorphism is observed along the active tectonic subduction zones of ocean-continent subduction (e.g. circum pacific belt,), ocean-ocean subduction (e.g. Japanese islands) and continent-continent subduction (e.g. Himalayan). Such type of metamorphism occurs on a large scale of a few 100 to 1000 sq.kms, and hence termed as regional metamorphism. In the earlier days, it is termed as regional dynamo-thermal metamorphism.

In regional metamorphism, the basin sediments accumulated in the fore-arc basin are gradually dragged and buried into the subduction zone, where both temperature and pressure are increased (Figs. 2.4 and 2.5). Temperature increase may be due to the heat generated by (a) burial at depth, (b) hot mantle convection, and (c) magma generation and crystallization at or near the vicinity of the subduction zone. Pressure increase is due to overloading of buried sediments at depth. However, temperature (T) and pressure (P) increase may not be uniform in all types of subduction zones. It is quite variable from one subduction to another subduction zone due to their angle and rate of subduction. In shallow subduction zones, P-T is generally ~150 to 200°C to a maximum of about 500° to 600°C and 5 to 6 kb, whereas in steep subduction zones it goes to a maximum of ~700° to 800°C and 8 to 10 kb., e.g. the lesser (LHC) and higher Himalayn crystalline (HHC) belts show low to high grade metamorphism (Fig. 3.2).

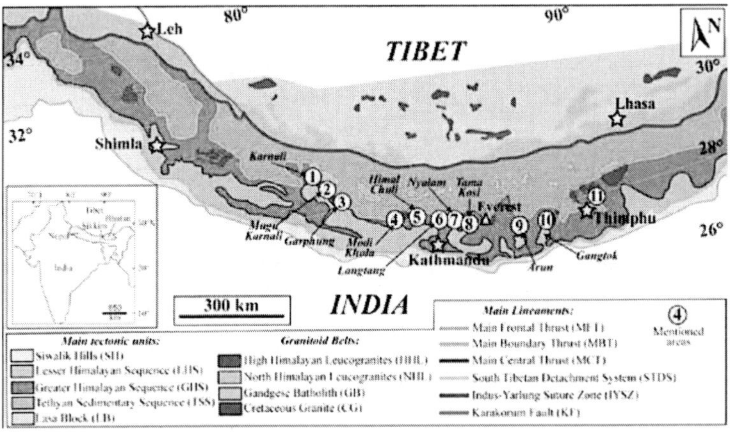

Fig. 3.2: Geological Map of the Himalaya (taken from the net by Montomoli et al., 2014, Geological Society of London Special Publication SP412.3)

Dark green – Greater Himalaya – medium to high grade metamorphic rocks; Light green – Lesser Hinalaya – diagenesis to medium grade metamorphic rocks; Metamorphic belt stretches ~2500 kms in length; ~100 kms in width

Chapter 4

Metamorphic Studies

The understanding of metamorphic process took place slowly than the formation of igneous and sedimentary rocks. It is due to the fact that igneous and sedimentary processes are observed on the earth's surface. Igneous process is observed from the volcanic activities that take place on the surface of the earth. It gives an idea about magma composition and its crystallization behavior. Mechanical sedimentation and chemical precipitation that take place in the basins of the earth surface give ample evidences to understand the sedimentation processes. However, metamorphism takes place beneath the surface of the earth at a depth of ~5 to ~50 kms; its process cannot be observed directly. It can be inferred only from the product rocks that show new set of minerals and structures.

4.1 Prograde Metamorphic Zones

The first observation that the rocks have undergone metamorphism is from the study in the north east Highland of Scotland by GM Barrow (1893, 1912). The map for this region is shown in Fig. 4.1. In the map, southern side is marked by highland boundary fault, followed towards north by chlorite zone, biotite zone (not indicated), garnet zone, staurolite zone, kyanite zone, sillimanite zone and granite.

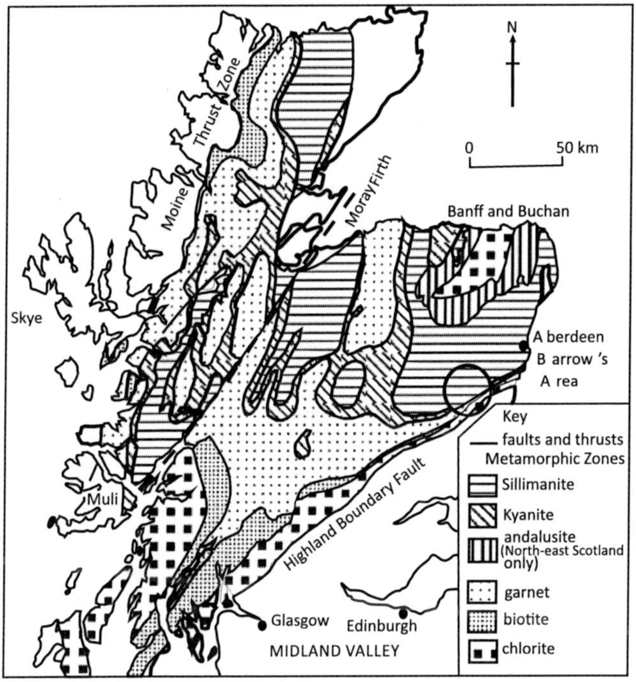

Fig. 4.1: Barrovian Zone Metamorphism of Scotland

In this region, the rocks are mainly pelites, some basic rocks and quartz-rich sandstone. The pelites have shown the development of new minerals, when going from lower to higher temperature side. Based on the appearance of new minerals, Barrow suggested a series of metamorphic zones. He observed that metamorphism becomes more intense while going from lower to higher side. The mineralogical changes are accompanied by grain size changes and their overall structure. Slate is found at the lower part, schist at the middle part and finally the gneiss at the higher part of that section. Depending on the appearance of new minerals from lower to higher part, Tilley (1925) coined the following zones:

- Chlorite zone
- Biotite zone
- Garnet zone
- Staurolite zone

- Kyanite zone
- Sillimanite zone

Tilley termed the zone boundaries as isograds. It means that all along the boundary between one to another zone, there will be appearance of new minerals called as index minerals, and disappearance of certain minerals. Once the new mineral is formed, it may persist into the middle and higher zones or may disappear in the higher zone. If the composition of the rock is not suitable like sandstone, there may not be any new minerals, and therefore, the isograd boundaries cannot be delineated. So the composition of the parent rock is important to demarcate different zones and isograd boundaries between the zones.

Since the observation was first made by Barrow, such metamorphic zones are termed as Barrovian Metamorphic Zones. Barrow's observation helped to study similar type of metamorphosed sediments in other parts of the world, and later found that Barrovian zone metamorphism is common in nature.

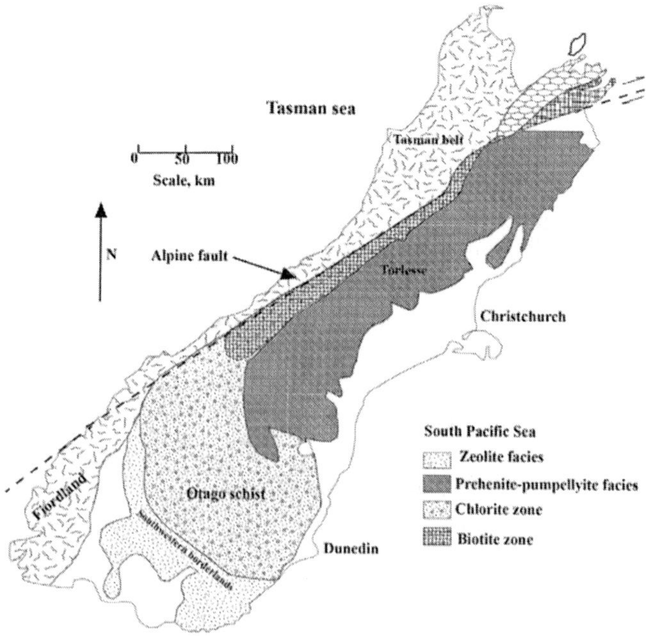

Fig. 4.2: Volcanic Rocks in New Zealand Showing Diagenesis to Low Grade Metamorphism

Burial metamorphism has been understood by the studies carried out in New Zealand. In this region, the rocks are mainly of volcanic origin and consist of greywackes and tuffs that deposited in deep trough and metamorphosed due to burial. This area was studied by D.S. Coombs and others in 1950 to 1960s and the map of that region is given in Fig. 4.2.

The southern part shows very low-grade metamorphism and the northern part medium Barrovian metamorphism. The different zones/isograds from the southern part towards north are:

1. Zeolite facies
2. Prehnite-Pumpellyite facies
3. Pumpellyite (-actinolite)
4. Greenschist facies

   Chlorite (-clinozoisite) zone

   Biotite zone

   Almandine (garnet) zone
5. Beginning of amphibolite facies

   Oligoclase (albite is replaced by a more calcic plagioclase)

The zeolite facies is developed only in the southern part of the area. In the lowest grade, the diagnostic minerals are developed in the very fine grained glassy matrix. The assemblages have shown incomplete reactions and metastable minerals. Along with some new minerals of metamorphism, pre-existing parent rock minerals are common. The low-grade rocks of southern New Zealand represent transitional zone from diagenesis to low-grade metamorphism. It is a classic area to study and understand the beginning of metamorphism.

Another type of metamorphism called contact metamorphism was first studied near Glinsk in Ireland by W.M. Edmunds and later by Pitcher and Burger in 1972. They observed that the granite pluton intruded the pelitic rocks of the Dalradian Super Group which were affected by contact metamorphism. Prior to the intrusion of granite, this area was affected by regional metamorphism and reached to garnet grade of greenschist facies conditions. Subsequently, due to granite intrusion new minerals biotite, andalusite, cordierite, k-feldspar and sillimanite are formed in this area. Very close to the intrusive contact, high temperature mineral sillimanite is observed and biotite farther away from the contact zone. These minerals appear as spots throughout the contact zone. The pelitic rocks show development of metamorphic aureole surrounding the granitic intrusion. This type of contact metamorphism is seen in the many parts of the world. However, the effect of contact metamorphism depends on the size and composition of the pluton and also the type of country rock involved. If the country rock is igneous, the intrusion will not have any effect on the country rock. If it is a sedimentary country rock, there may be effect of intrusion depending upon the composition of the country rock. If the size of the pluton is very small, there may be little effect of contact metamorphism. In case of large pluton, the contact metamorphism may be significant with the development of metamorphic aureoles. If the composition of the pluton is of ultramafic or mafic, contact metamorphic effect is insignificant even with suitable country rock. This is due to the presence of low volatiles in such plutons. Only acidic pluton like granodirite, granite can develop contact metamorphism with suitable country rock compositions like shale, impure limestone, and grewacke, etc. The basic pluton has very high temperature of ~1200°C and less H_2O (< 0.1%), there is no contact metamorphism. However, the granitic plutons with temperature of 700-900°C and high volatiles especially H_2O (~2% to 8%) show significant contact metamorphism. The volatiles present in the plutons transmit heat from the pluton to the country rock very easily and cause contact metamorphism.

4.2 Paired Metamorphic Belt

Paired metamorphic belt was first suggested by Miyashiro in 1961 from his study of Shikoku Honshu islands of Japan. In this region, paired metamorphic belt is exposed along the north eastern-southwestern access that occur parallel to the active subduction zone. The inner north western belt, called Ryoke or Abukuma belt has high temperature-low pressure metamorphic rocks. The outer belt is called as the Sanbagawa belt shows high pressure-low temperature metamorphic rock (Fig. 4.3).

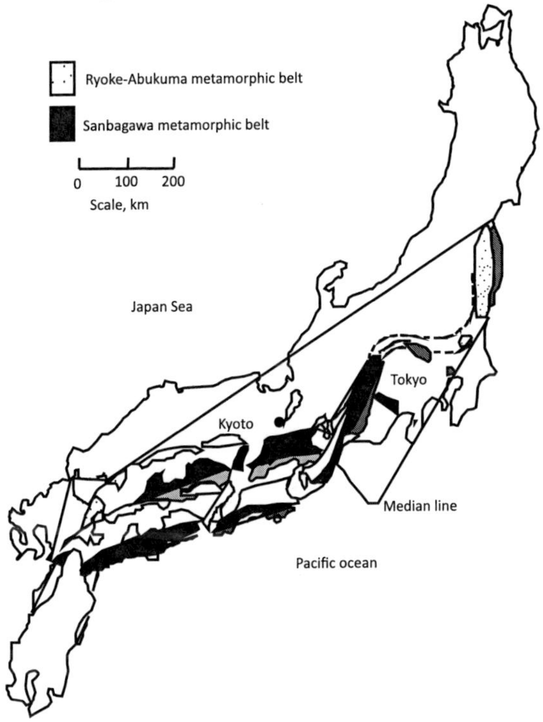

Fig. 4.3: Paired Metamorphic Belt—Ryoke or Abukuma Belt and Sanbagawa Belt in Japan

These two belts are separated along the whole length by a major fault called a median line. The Ryoke-Abukuma belt consists of mature volcanic arc sediments. The Sanbagawa belt consists of arc derived sediments and the mixture of volcanic oceanic crust and marine sediment. Subsequent to Miyashiro's study, many paired metamorphic belts are reported in many parts of the world.

Chapter 5

Equilibrium Mineral Assemblages or Mineral Paragenesis

A rock containing stable mineral assemblage is considered to be at thermodynamic equilibrium under specific temperature, pressure and composition. At equilibrium the composition of each mineral is fixed. The mineral assemblages, that are at equilibrium in the given P-T condition, is known as mineral parageneses (Fig. 5.1).

In metamorphic terrains, it is difficult to demonstrate that the mineral assemblages are at thermodynamic equilibrium once the rock has formed. Only from the following indirect evidences, one may be able to conclude that the mineral assemblages represent thermodynamic equilibrium.

(1) There is no reaction or disequilibrium texture such as replacement texture, reaction corona, compositional zoning, etc.

(2) Each mineral is in physical contact with every other mineral and have grain boundary network in the rock under consideration (Fig. 5.1).

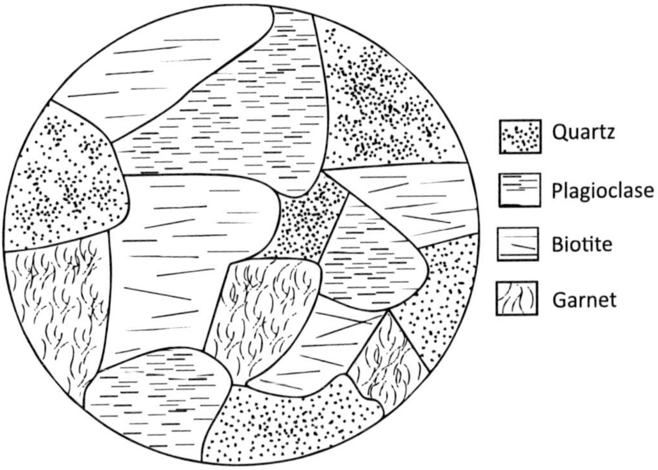

Fig. 5.1: Mineral Parageneis: Qtz-Plag-Bt, Grt-Bt-Plag, Qtz-Grt-Bt Three Phases Coexist at Thermodynamic Equilibrium

(3) Non-layered rocks are likely to be at equilibrium than the layered rocks. In the layered rocks, the minerals in one layer may not be at equilibrium with the minerals occurring in the next layer. However, minerals in each layer may be at equilibrium.

(4) Textural equilibrium is difficult to achieve than the chemical equilibrium.

(5) Mineral assemblages confirm phase rule for the given rock compositions. If the rock has more number of phases than the number of components, it is suspected to be disequilibrium assemblage. If the number of component is 4, as per phase rule, maximum number of phases is 6 considering only 2 variables temperature and pressure. If the phases are exceeding 6, the mineral assemblage may be at disequilibrium condition.

(6) Analyzing the composition of each mineral from core to rim gives information about the presence of uniform or variable compositions. If all the minerals show uniform chemical composition, it suggests chemical equilibrium. Presence of zoned minerals suggests that the rock is in disequilibrium condition.

On many occasions, a suite of metamorphic rocks in the field provide evidence for thermodynamic equilibrium. For example, if the rocks of similar composition in an area show similar mineral assemblages, it implies that they were equilibrated at the same P-T condition. If rocks demonstrate systematic changes in mineral assemblage over larger distances, it implies that systematic equilibration has taken place with mineralogical changes along the thermal gradient.

In order to understand mineral parageneses, certain basic rules are necessarily to be understood. They are described here.

5.1 Phase Rule and Phase Equilibrium

To understand the phase rule and phase equilibrium, certain terms like system, phase, component, variables, etc., are to be understood.

5.1.1 System

It is described as some portion of the universe that will be studied. For example, a rock on the surface of the earth is surrounded by the atmosphere. In this case, rock is one system and the surrounding atmosphere is another system. In surface of the earth, i.e. in the field, an igneous rock is associated with the sedimentary rock in which the former forms one system and the latter forms another system.

5.1.2 State of a System

In the system, the matter may be solid, liquid, gas or combination of any these physical state. The state of the system means that it is a measure of specific temperature, pressure, volume, mass, density and composition of matter.

As an example: Water.

Liquid water is stable at 25°C temperature, one bar pressure and with fixed chemical composition (H_2O).

Solid water or ice is stable < 0.0°C, one bar pressure and with fixed chemical composition (H_2O).

Water vapour or steam forms >100°C, one bar pressure and with fixed chemical composition (H_2O).

5.1.3 Open System

In an open system, the molecular elements/components, volatiles, heat, etc. are exchanged between one system to another system. For example, when the granite melt intrudes into sedimentary country rock such as carbonates, sandstone etc., heat, volatiles, molecular elements/components, etc. are exchanged between igneous and sedimentary rock producing new minerals. Since the matter is exchanged between two rocks, the system is called an open system.

5.1.4 Closed System

In closed system, the molecular elements/components, volatiles, heat, etc. are not exchanged between one system to another system. In the sedimentary rocks, if we have alternate layers of shale and greywacke, both the layers undergo mineralogical changes independently with change in physic-chemical condition but no exchange of components between the layers. In other words, shale produces a set of new minerals within its layer independently and greywacke produces another set of new minerals in its layer separately during metamorphism. In the process, the matter is not exchanged between the layers of shale and grewacke, except heat or sometimes fluids. So each layer of rock behaves like a closed system.

5.2 Phase

Phase is a definite physical entity of a system that is mechanically separable. A phase may be a mineral, liquid, gas or an amorphous solid. For example, ice is a solid phase, water is a liquid phase, and steam is a gaseous phase.

5.3 Component

The chemical composition of a phase can be defined in terms of elements, molecules or end member components.

***Example* 1**

The composition of garnet is written as:

$(Mg_{1.4}, Fe_{1.9}, Mn_{0.1}, Ca_{0.6})_3\ Al_2\ Si_3\ O_{12}$

It is a mixture of four minerals Pyrope, Almandine, Grossularite and Spessartine.

Their compositions are:

Pyrope (Pyp)	$Mg_3\ Al_2\ Si_3O_{12}$	or	$3MgO\ Al_2\ O_3\ 3\ SiO_2$
Almandine (Alm)	$Fe_3\ Al_2\ Si_3O_{12}$	or	$3FeO\ Al_2\ O_3\ 3\ SiO_2$
Grossularite (Grs)	$Ca_3\ Al_2\ Si_3O_{12}$	or	$3CaO\ Al_2\ O_3\ 3\ SiO_2$
Spessartine (Spes)	$Mn_3\ Al_2\ Si_3O_{12}$	or	$3MnO\ Al_2O_3\ 3\ SiO_2$

It can be defined in terms of seven elements—Mg, Fe, Mn, Ca, Al, Si and O

or

in terms of six molecules—MgO, FeO, MnO,CaO, $Al_2\ O_3$, SiO_2

or

in terms of four end-member components—

$Mg_3\ Al_2\ Si_3O_{12}$ (Pyp), $Fe_3\ Al_2\ Si_3O_{12}$ (Alm), $Ca_3\ Al_2\ Si_3O_{12}$ (Grs), $Mn_3\ Al_2\ Si_3O_{12}$ (Spes)

As per phase rule, the minerals should be defined only with the minimum number of components. In the above example, garnet can be defined by four end-member components, rather than by seven element components or by six molecular components.

Example 2

Potash feldspar composition is as follows: $(Na_{0.1}, K_{0.85}, Ca_{0.05})_1\ Al_1\ Si_3\ O_8$.

It is a mixture of potash feldspar, albite and anorthite minerals.

Potash Feldspar	$KAlSi_3O_8$	or	$K_2O\ Al_2O_3, 6SiO_2$ (KF)
Albite	$NaAlSi_3O_8$	or	$Na_2O\ Al_2\ O_3\ 6SiO_2$ (Ab)
Anorthite	$CaAl_2Si_2O_8$	or	$CaO\ Al_2O_3\ 2SiO_2$ (An)

Potash feldspar can be defined using six elements—K, Na, Ca, Al, Si and O or as five molecular components—K_2O, Na_2O, CaO, Al_2O_3 and SiO_2 or as three end-member components—$KAlSi_3O_8$ (KF), $NaAlSi_3O_8$(Ab), $CaAl_2Si_2O_8$(An). It is possible to define by three end-member components, rather than by six elements or five molecular components.

5.4 Variables

There are two types of variables, namely the extensive and the intensive variables. *Extensive variables* depend on the quantity of the material including mass, volume and number of molecules in the system. The *intensive variables* are temperature, pressure, density, etc. and they are independent of the size, mass and volume of the system.

If we divide one extensive variable by another extensive variable, the ratio is the intensive variable of the system.

For example: Mass/Volume; volume of a phase/number of moles of a phase-produce intensive variables.

5.5 Prograde Metamorphism

It refers to an increase in metamorphic grade with temperature and pressure, due to which mineralogical and structural changes occur in the rock being subjected to metamorphism.

5.6 Retrograde Metamorphism

It means that the grade of the metamorphosed rock is decreased during exhumation and cooling. Any mineralogical changes occurred during cooling is known as retrograde metamorphism.

Progressive metamorphism means that the highest grade metamorphic rocks is progressed through a sequence of mineral assemblages from low P-T to high P-T conditions, maintaining equilibrium at every stage. In other words, the high-grade metamorphic rocks progressed slowly with time from low, medium to high-grade conditions, without going directly from low to high-grade conditions. The strong evidence for progressive metamorphism comes from textural studies. The metastable or unreacted minerals of the low grade are found as relicts or inclusions in high-grade mineral assemblages.

Chapter 6

Importance of Rock Compositions

Primary rock compositions play a significant role in understanding the metamorphic process. Metamorphism takes place in a variety of tectonic settings with different geothermal gradients, overload and directed pressures. The rock compositions and tectonic settings that produce metamorphic rocks may be different from one terrain to another terrain.

As mentioned earlier, the rate and angle of subduction are quite different in various subduction zones. In low angle subduction zones, very low to medium-grade metamorphic facies, steep subduction zones medium to high pressure and temperature metamorphic facies, and in moderate subduction zones low to high-grade metamorphic facies may be present. Though the tectonic setting is important, the composition of the country rock is equally important to understand metamorphism through mineral assemblages. Mineral assemblages help in understanding P-T condition of metamorphism. It may be pointed out that sedimentary rocks are most affected as compared to igneous rocks by metamorphism.

6.1 Types of Parent Rock or Protolith

The whole rock chemical composition and mineralogy of the parent rock is important to understand the mineralogical changes that occur during metamorphism. When we study metamorphic rocks, it is important to keep "parental rock" mineral composition in mind. The mineral assemblage of the metamorphic rock gives information about parent rock mineral composition.

Igneous rocks are not easily affected by metamorphism, as they are crystallized at higher temperature (900-1500°C) than temperature (200-800°C) of metamorphism. Igneous rock minerals crystallize at higher temperature than the minerals of metamorphism. However, if the igneous rock is altered or weathered, the primary minerals are converted to low temperature minerals such as serpentine, clays, sericite, etc. When the weathered rock undergoes metamorphism, different metamorphic minerals are produced, depending on the P-T condition of metamorphism. The different types of igneous rocks are described briefly in the following section.

6.2 Igneous Rocks

Various types of igneous rocks, their mineralogy and whole rock chemical composition are described here in some detail.

Ultramafic rocks—They are mantle derived rocks that include komatiites, cumulates, peridotite, pyroxenite and dunite. These rocks are high in Mg, Fe, Ni, and Cr and poorer in Si, Al, Na and K. The main minerals are olivine, pyroxene and Ca rich plagioclase. These rocks are not directly affected by metamorphism. They are weathered or hydrothermally altered to serpentine-chlorite-zeolite bearing rocks. Such altered rocks are affected by metamorphism.

Mafic rocks—Basalts, gabbros and some greywacke's. Main minerals include olivine, pyroxene, amphiboles and Ca rich plagioclase. They are high in Fe, Mg, Ca components and poorer in Na, K, and Al components. They are hydrothermally weathered or altered to serpentine-chlorite-zeolite rocks. Such altered rocks can be affected and converted to chlorite schist, greenschist and amphibolite by metamorphism.

Intermediate rocks—Different varieties of diorite and their volcanic equivalents are known as intermediate rocks. They contain Ca-plagioclase, amphiboles and little pyroxene and quartz. They have higher amount of Ca, moderate amounts of Si, Al, Na, K, and poorer in Mg and Fe. These rocks are altered to clay matter, and affected by metamorphism.

Acidic rocks. They are primarily granodiorites- granites and their volcanic equivalents. They consists mainly K-feldspar, Na-rich plagioclase, quartz, mica and little amount of amphibole and biotite. They are easily affected by weathering and alteration producing clayey rock.

Such clayey rocks are converted to phyllite, mica schist and gneiss due to metamorphism.

6.3 Sedimentary Rocks

Sedimentary rocks undergoing metamorphism may have the following lithologies:

1. Sandstone – mainly quartz with some or no feldspars
2. Limestone – carbonate minerals like calcite, dolomite, siderite, rhodochrosite, etc.
3. Greywacke – clay minerals, glauconite, some mafic minerals, some carbonates
4. Shale only – clay minerals illite, montmorillonite, kaolinite, etc.
5. Sandstone + shale in different proportions
6. Limestone + shale in different proportions
7. greywacke + shale in different proportions
8. Volcanic rocks + shale in different proportions
9. Volcanic rock + shale + limestone + sandstone in different proportions
10. greywacke + shale + limestone + sandstone in different proportions

Besides these combinations of lithologies, even individual rock like shale may have impurities of carbonates, iron oxides, sulphates or other compositions. Similarly, sandstone may have impurities of clays, carbonates and iron oxides. Therefore, these rocks produce different mineral assemblages in each facies condition. Therefore, to decipher the metamorphic facies or grade of an area, the primary mineral assemblages of various rocks are to be considered.

7.2 Discontinuous Reaction

This reaction is a net transfer reaction, where the reacting phases, in general, are pure components. They react at a particular P-T condition and the reaction is univariant. In such reaction, the reactant minerals are totally consumed and new product minerals are formed once the reaction or isograd boundary (P-T) is crossed. In Fig.7.2, the following two discontinuous univariant reactions are shown.

Muscovite + Quartz = Kfeldspar + Aluminosilicate + H_2O (dehydration reaction)

Calcite + quartz = wollastonite + CO_2 (decarbonation reaction)

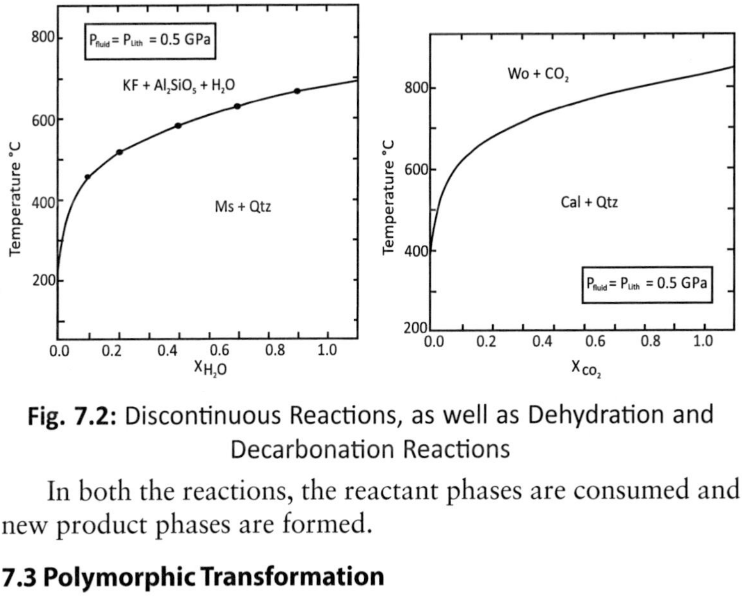

Fig. 7.2: Discontinuous Reactions, as well as Dehydration and Decarbonation Reactions

In both the reactions, the reactant phases are consumed and new product phases are formed.

7.3 Polymorphic Transformation

It is a kind of net transfer reaction. Single component polymorphic transformations are the simplest reaction to deal with. For example, SiO_2 is a single component. SiO_2 at low P-T condition occurs in the form of quartz. If quartz is subjected to high P-T condition, SiO_2 composition remains the same, except for the rearrangement of internal crystal structure. Quartz is converted to cristobalite or tridymite at high temperatures; or coesite or stisohvite at high pressures. All these minerals are

SiO_2 in composition. Similarly, graphite (C) is converted to diamond at very high pressure (C). Calcite ($CaCO_3$) is converted to aragonite ($CaCO_3$) at high pressure. Aluminosilicate (Al_2SiO_5) occurs; at low pressure andalusite, high pressure kyanite, high temperature sillimanite and at very high temperature mullite which is rare in occurrance. Only the first three polymorphs are commonly found (Fig. 7.3). The minerals having same chemical composition but with different atomic structures are called as polymorphs. The conversion of one polymorph to another polymorph takes place at specific P-T conditions along a univariant line and is true for other polymorphs.

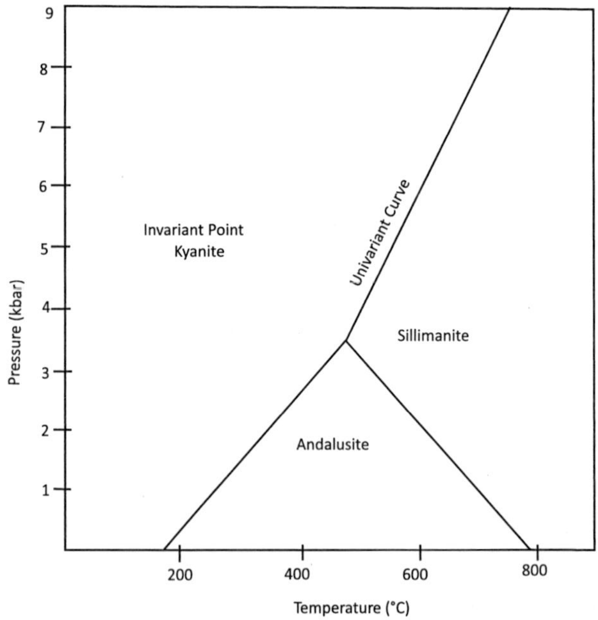

Fig. 7.3: Al_2SiO_5 Polymorphs with Invariant Point and Univariant lines

7.4 Solid-Solid Net Transfer Reaction

It is another kind of net transfer reaction. It involves solids of different compositions where solids on the reactant side are totally consumed and new minerals are formed on the product side.

Among the sedimentary rocks, sandstone and pure limestone are not considered as they do not produce any new metamorphic minerals. They are recrystallized, i.e. sandstone to quartzite; limestone to marble. However, volcanics/greywackes of basic composition, argillaceous and impure carbonates compositions undergo lot of mineralogical changes during metamorphism. Argillaceous and basic rock compositions give enough information to decipher the facies/grades of metamorphism. Basic rock compositions show lot of mineralogical changes at very low to low-grade metamorphism. They are sensitive to pressure and give information about low, medium and high pressures at very low to low temperature conditions. Argillaceous rocks do not show significant mineralogical changes at very low-grade metamorphism. However, it shows mineralogical changes from low grade to very high-grade conditions. Therefore, if the terrain contains these two rocks, one can decipher the prograde/retrograde condition of metamorphism. If only argillaceous rock is present in a metamorphic terrain, very low-grade metamorphism cannot be ascertained. Similarly, if basic rocks are present in the metamorphic terrain, it does not give any clear information about medium to high-grade metamorphism. Beside these two compositions, other compositions like marls, sandstone, limestone, quartzo-feldpathic rocks, etc. may give some information about metamorphism, but not useful to decipher the metamorphic facies/grades of an area. In view of the above-mentioned aspect, sedimentary rocks mineralogy and their whole rock chemical compositions is briefly described.

Shales and mudstones (Pelites)—They are one of the most common sedimentary rocks present on the earth's surfaces. They are fine grained having clastic clays, muds, silts and sands. They are high in Al, K, Si, H_2O and poorer in Mg, Fe and Ca. These sediments have clay minerals such as kaolinite, illite, montmorillonite, smectite, etc., detrital muscovite, quartz, albitic plagioclase, K-feldspar, iron oxide. Sometimes, they contain carbonaceous minerals such as calcite, dolomite, siderite, limonite, organic carbon, etc. as impurities termed as carbonaceous shale. If iron oxide is more in the shale, it is called as ferruginous shale. Carbonates (calcareous) are mostly limestone and dolostone, high in Ca, Mg and CO_2. The minerals include calcite and dolomite. Impure carbonates (marls) may contain sandstone or shale components.

Quartz rich rocks—It includes cherts of oceanic and sandstone of continental clastics. It is mainly having SiO_2. The minerals are varieties of amorphous silica and quartz.

Quartzo-feldspathic rocks—They include quartzo-feldspathic arkose, breccia, conglomerate, granitoid and rhyolitic rocks. They are high in Si, Na, K, Al and poorer in Mg and Fe. The rocks contain mainly quartz, potash feldspar, plagioclase, muscovite, biotite, chlorite and some amphiboles.

Chapter 7

Metamorphic Reactions

As described earlier, metamorphism involves re-crystallization and formation of new minerals from the parent rock in solid state through the process of diffusion. The changes are mainly due to many mineral reactions taking place during metamorphism that produce new minerals from the existing ones. Several types of reactions take place during metamorphism. The reactions proceed in steps when going from lower-middle, middle-higher and higher-very high P-T conditions. The reactions discussed here may take place independently or simultaneously in a rock that depends on the mineralogy of the rock undergoing metamorphism.

The following metamorphic reactions are described based on the research done by many workers. Their work is based on field, microscopic and other laboratory investigations done on various metamorphic rocks. When the rock is undergoing metamorphism, it shows changes in mineralogy and structure. The most noticeable change is the formation of new minerals by metamorphic reactions. The line demarcating the appearance of new mineral is called as an isograd. The reactions indicate the appearance or disappearance of the minerals in a rock due to change in P-T condition. This helps in understanding the process of metamorphism, delineation of isograds, and the physical and chemical conditions responsible for such changes. Many mineral reactions are given based on various studies that are briefly described here.

The reactions are broadly classified into discontinuous and continuous reactions. However, the reactions may be further classified into (i) polymorphic transformations, (ii) solid-solid net transfer reactions, (iii) ex-solution reactions, (iv) de-volatilization reactions, (v) ion-exchange reactions, (vi) oxidation-reduction reactions and (vii) reaction involving dissolved species.

7.1 Continuous Reaction

In continuous reaction, the reacting phases are solid solutions. The reaction starts at lower P-T and proceeds towards higher P-T but such reaction is limited to a few metamorphic zones. For example, the reaction Chlorite + Quartz = Garnet + Water.

In this reaction (Fig. 7.1), if chlorite is pure iron end member, the reaction will proceed at lower temperature. Alternatively, if chlorite is pure Mg end member, the reaction will takes place at higher temperature. However, the natural chlorite is solid solution between Mg and Fe end members. So the garnet forming reaction starts at low temperature with Mg/MgFe ratio of chlorite of 0.3 and continues towards the higher ratio of 0.8 at higher temperature. As the reaction starts at low temperature and continues to higher temperature, this type of reaction is known as continuous reaction.

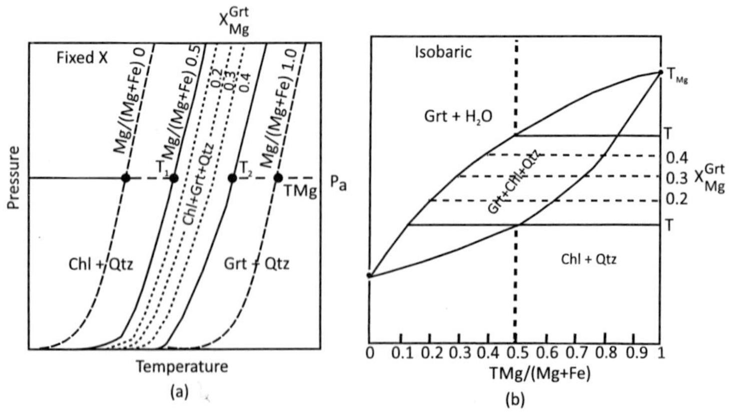

Fig. 7.1: Continuous Reaction with Different Moles of Mg in Garnet
(a) with fixed composition and variable temperature
(b) with fixed pressure

7.2 Discontinuous Reaction

This reaction is a net transfer reaction, where the reacting phases, in general, are pure components. They react at a particular P-T condition and the reaction is univariant. In such reaction, the reactant minerals are totally consumed and new product minerals are formed once the reaction or isograd boundary (P-T) is crossed. In Fig.7.2, the following two discontinuous univariant reactions are shown.

Muscovite + Quartz = Kfeldspar + Aluminosilicate + H_2O (dehydration reaction)

Calcite + quartz = wollastonite + CO_2 (decarbonation reaction)

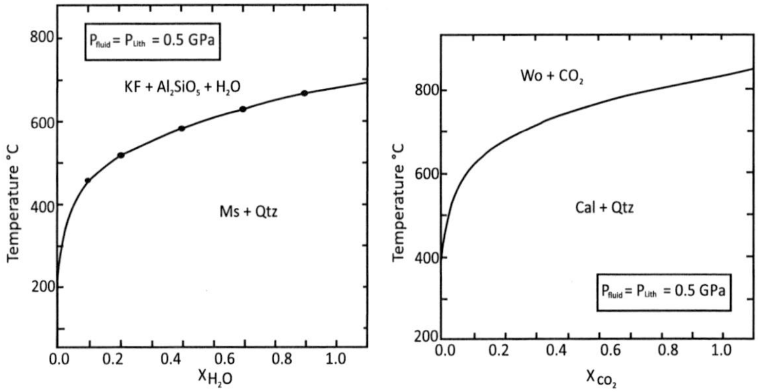

Fig. 7.2: Discontinuous Reactions, as well as Dehydration and Decarbonation Reactions

In both the reactions, the reactant phases are consumed and new product phases are formed.

7.3 Polymorphic Transformation

It is a kind of net transfer reaction. Single component polymorphic transformations are the simplest reaction to deal with. For example, SiO_2 is a single component. SiO_2 at low P-T condition occurs in the form of quartz. If quartz is subjected to high P-T condition, SiO_2 composition remains the same, except for the rearrangement of internal crystal structure. Quartz is converted to cristobalite or tridymite at high temperatures; or coesite or stisohvite at high pressures. All these minerals are

SiO_2 in composition. Similarly, graphite (C) is converted to diamond at very high pressure (C). Calcite ($CaCO_3$) is converted to aragonite ($CaCO_3$) at high pressure. Aluminosilicate (Al_2SiO_5) occurs; at low pressure andalusite, high pressure kyanite, high temperature sillimanite and at very high temperature mullite which is rare in occurrance. Only the first three polymorphs are commonly found (Fig. 7.3). The minerals having same chemical composition but with different atomic structures are called as polymorphs. The conversion of one polymorph to another polymorph takes place at specific P-T conditions along a univariant line and is true for other polymorphs.

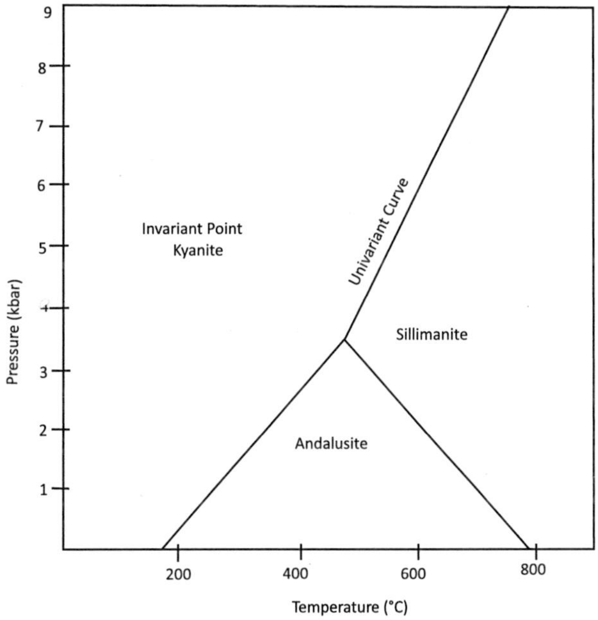

Fig. 7.3: Al_2SiO_5 Polymorphs with Invariant Point and Univariant lines

7.4 Solid-Solid Net Transfer Reaction

It is another kind of net transfer reaction. It involves solids of different compositions where solids on the reactant side are totally consumed and new minerals are formed on the product side.

Examples

Jadeite + quartz = albite

$(NaAlSi_2O_6 + (SiO_2) = (NaAlSi_3O_8)$

Enstatite + anorthite = Diopside + sillimanite

$(MgSiO_3 + CaAl_2Si_2O_8) = (CaMgSi_2O_6 + Al_2SiO_5)$

Orthopyroxene + plagioclase = Garnet + clinopyroxene + quartz

$((Mg, Fe)SiO_3) + (CaAl_2Si_2O_8) = (Mg, Fe)_3Al_2Si_3O_{12}) + (Ca(Mg, Fe)Si_2O_6) + (SiO_2)$

In another type of net transfer reaction, the reactant minerals may have volatiles which are totally conserved in the product minerals as given below:

Talc + enstatite = anthophyllite

$(Mg_3Si_4O_{10}(OH)_2 + 4MgSiO_3) = (Mg_7Si_8O_{22}(OH)_2)$

Sometimes, in net transfer reactions the diffusion may be very slow. In such case, the reaction will not go to completion, instead reaction corona (rim) is developed between the reactant and the product, arresting further reaction. When solid solution is limited in the mineral, the net transfer reaction is discontinuous and univariant (Fig. 7.2), and the reaction goes to completion on reaching a particular grade.

7.5 De-volatilization Reactions

The hydrous and carbonate minerals involved in metamorphic reactions evolve H_2O or CO_2 or other volatile such as SO_2, Cl, CH_4. Such volatile releasing reactions are known as de-volatilization reactions. If it is releasing H_2O, it is known as de-hydration reaction (Fig. 7.4). If it is CO_2, it is known as de-carbonation reaction (Fig. 7.4).

De-hydration reaction (a)

Muscovite + quartz = Kfeldspar + aluminosilicate + H_2O

$(KAl_2Si_3AlO_{10}(OH)_2 + SiO_2) = (KAlSi_3O_8 + Al_2SiO_5 + H_2O)$

De-carbonation reaction (b)

Calcite + quartz = wollastonite + CO_2

$(CaCO_3 + SiO_2) = (CaSiO_3 + CO_2)$

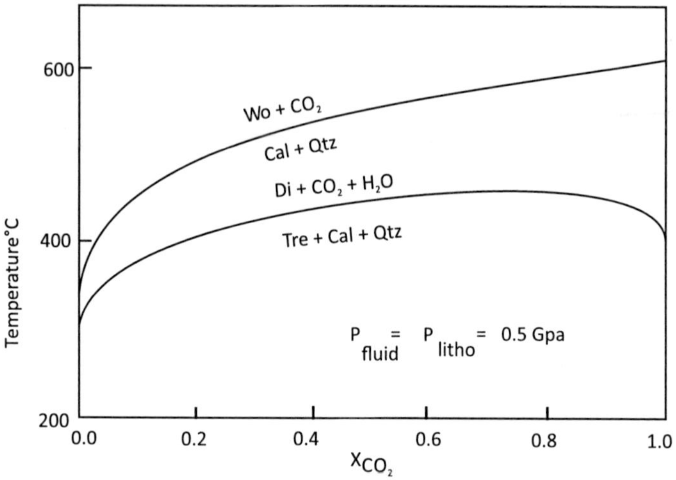

Fig. 7.4: Discontinuous De-volatilization Reaction

7.6 Discontinuous De-volatilization Reaction

The de-volatilization reactions take place not only with increasing temperature, but also with partial pressure of volatiles. Rock is usually saturated with H_2O and/or CO_2 but significantly less in other volatiles and are not so important in metamorphic reactions. The partial pressure of volatiles is defined as:

$P_{H_2O} = P_{litho} = P_f = 1; P_{H_2O+CO_2} = P_{litho} = P_f = 1$

Here the partial pressure of pure water is equal to 1, similarly partial pressure CO_2 is equal to 1.

(H_2O – water pressure; litho – lithostatic pressure; f – fluid pressure)

In case of mixed volatiles, it is given as

$P_{H_2O+} P_{CO_2} = P_{litho} = P_f = 1$

Generally, the rock contains H_2O + CO_2 in varying proportions but the total of these two is equal to 1.0. The H_2O and CO_2 ratios are expressed by the following methods:

H_2O = 0.6 moles : CO_2 = 0.4 moles

H_2O = 0.68 moles : CO_2 = 0.32 moles

H_2O = 0.3 moles : CO_2 = 0.7 moles

H_2O = 0.8 moles : CO_2 = 0.2 moles

So in all cases total P_f is equal to 1 which is equal to P_{litho}. The amount of pressure exerted either by H_2O or CO_2 is known as partial pressure of H_2O or CO_2. It is written as P_{H_2O} or P_{CO_2} or as f_{H_2O} or f_{CO_2} for volatiles. (f = fugacity or vapour pressure)

The partial pressure of H_2O or CO_2 is calculated by the following method.

Examples: $H_2O + CO_2$ = 0.6 moles + 0.4 moles = 1

$$P_{CO_2}(f_{CO_2}) = \frac{0.4}{(0.6+0.4)=1} = 0.4$$

$H_2O + CO_2$ = 0.68 moles + 0.32 moles = 1

$$P_{H_2O}(f_{H_2O}) = \frac{0.68}{(0.68+0.32)=1} = 0.68$$

In these two cases, the partial pressure of CO_2 and H_2O is 0.40 and 0.68 respectively. It is less than lithostatic (P_{litho}) or total pressure (P_t).

Sometimes, the rocks at higher grades may not have saturated with volatiles (mainly H_2O), but some of the hydroxyl minerals involved in the reactions may release volatiles. Even in some cases, the rock may be less in volatiles giving rise to $P_{fluid} \neq P_{litho}$ situation. Therefore, the reaction in water saturated condition will proceed at lower temperature. In H_2O under saturated condition ($f_{H2O} < 1.0$) the reaction will proceed at higher temperatures depending on the f_{H_2O} of water released by one such reaction.

Muscovite + quartz = Kfeldspar + aluminosilicate + H_2O ($KAl_2Si_3AlO_{10}(OH)_2 + SiO_2$) = ($KAlSi_3O_8 + Al_2SiO_5 + H_2O$)

7.7 Exsolution reaction

Solid solution minerals have mixture of end-member components that are partially or completely mixed depending on the composition of whole rock and P-T condition. Example:

Complete solid solution minerals

Olivine = mixture of two end members – Fosterite (Mg_2SiO_4) + Fayalite (Fe_2SiO_4)

Garnet = mixture of four end members – Pyrope + Almandine + Grossularite + Spessartine

In these minerals, the end members may mix at any proportion and they depend on whole rock chemical composition and P-T condition. In the case of olivine, Mg rich whole rock composition will produce fosterite rich (Mg rich) olivine. If the rock is enriched in Fe composition, it will produce fayalite rich (Fe rich) olivine. Garnet also shows similar kind of complete solid solution. Such complete solid solution minerals do not show ex-solution on cooling.

Fig. 7.5 (a)

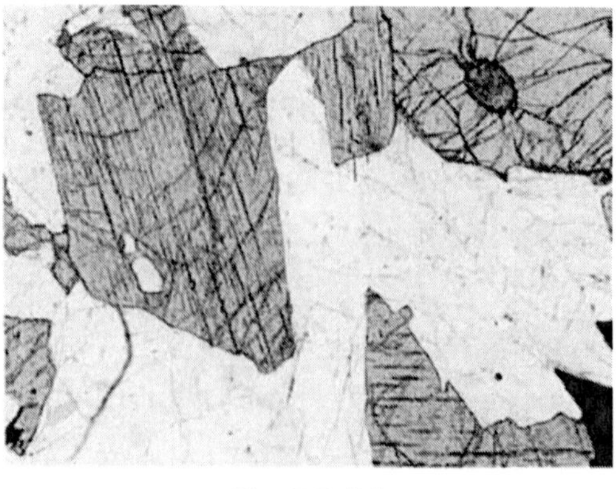

Fig. 7.5 (b)

Fig. 7.5: Exsolutions in Granulite (a) photomicrograph of plagioclase showing scapolite exsolution (grey showing elongated scapolite needles) (b) photomicrograph of clinopyroxene (pale greenish yellow colour) showing orthopyroxene exsolution along cleavage direction

However, many minerals like K-feldspar, pyroxenes, amphiboles, etc. have limited solid solutions. At high P-T condition, the end members may mix completely or partially. During cooling, some end members do not remain in the mineral, but comes out as a separate phase within that mineral. This phenomenon is known as exsolution. In K-feldspar (K)-albite (Na)-anorthite (Ca) end members are present as one phase at high temperature. But on cooling, anorthite end member comes out, because Ca is smaller than Na and K ions, as exsolved lamellae. Same thing is true for plagioclase, where K-feldspar comes out as exsolved lamellae on cooling. In the case of orthopyroxene, clinopyroxene (calcium end members) comes out as exsolved lamellae and vice-versa. This type of exsolutions are common in minerals with limited solid solution (Fig. 7.5).

7.8 Ion Exchange Reaction

In pelites and greywackes, certain minerals are present in all metamorphic grades. For example, in pelites garnet-biotite minerals are present from upper greenschist facies (garnet grade) to almandine-amphibolite facies (upper sillimanite grade) rocks. From lower to higher grades, garnet-biotite chemical composition change continuously. There are many such pairs like garnet-clinopyroxnene, garnet-chlorite, garnet-orthopyroxene, etc. For details of such pairs, see book by Winter, 2012. Such pair of minerals show different compositions in each grade. Garnet and biotite, each one may have one Fe/(Fe + Mg) ratio in the lower grade, that are changed to different ratios in medium grade, and change further in the higher grades. It suggests that such changes in Fe/(Fe + Mg) ratio towards higher grade is through exchange of elemental ions between the minerals and addition of components from the parent rock for their growth. If Fe is enriched in the garnet at low grade, biotite may relatively be enriched in Mg. In higher grade, garnet may become more enriched in Mg and biotite in Fe. Such exchange of components between minerals is known as ion exchange reaction and used to determine temperature of metamorphism.

As mentioned earlier, garnet and biotite are common in pelites of all metamorphic grades. The reaction is written as:

Garnet + Biotite = Garnet + Biotite,

This is rewritten as

$\text{Garnet}_{(\text{Pyrope(Mg)}}$), + $\text{Biotite}_{(\text{Annite(Fe)}}$), = $\text{Garnet}_{(\text{almandine(Fe)}}$), + $\text{Biotite}_{(\text{phlgopite(Mg)}}$

Similarly, in basic rocks, orthopyroxene and clinopyroxene are common at medium to higher grades.

The reaction is written as

Enstatite + Hedenbargite = Ferrosilite + Diopside. It is rewritten as

$\text{Orthopyroxene}_{(\text{Mg})}$ + $\text{Clinopyroxene}_{(\text{Fe})}$ = $\text{Orthopyroxene}_{(\text{Fe})}$ + $\text{Clinopyroxene}_{(\text{Mg})}$

Several types of such exchange reactions are observed in varieties of metamorphic rocks.

For each of those reactions, thermodynamic as well as experimental data are available. Using chemical composition analysed for the minerals of the exchange reaction, it is possible to determine temperature.

7.9 Oxidation-Reduction Reaction

Some minerals liberate oxygen during progressive metamorphism or consume oxygen during cooling.

In the reaction - hematite = magnetite + O_2.

O_2 is liberated, therefore, it is known as reduction reaction (Fig. 7.6).

In the second reaction – Magnetite + O_2 = hematite (Fig. 7.6).

O_2 is consumed, hence, it is known as oxidation reaction. Such reactions are common in the rocks having more iron rich minerals. Such reactions are known as oxidation-reduction reactions.

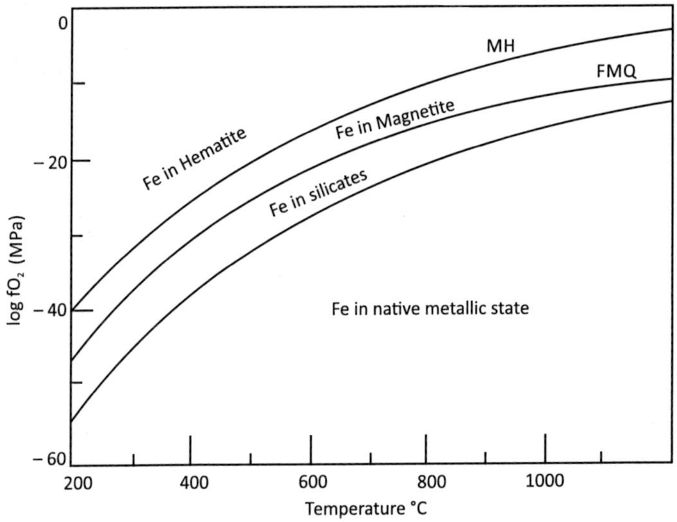

Fig. 7.6: Oxidation-Reduction Reaction MH—magnetite-hematite; FMQ—fayalite-magnetite-quartz

Chapter 8

Facies

The facies concept is conceived to understand metamorphism of the metamorphic terrain. The pioneering works of Goldschmidt (1911, 1913) and Eskola (1914, 1915) gave an idea about the mineral assemblages and their relation to chemical composition of the metamorphic rocks. Goldschmidt studied contact metamorphism of pelites, psammites, impure calcareous sediments, that occur surrounding a granitic pluton in Oslo region, southern Norway. He found that rocks of different chemical composition have produced three to five new minerals by contact metamorphism. He was the first one to suggest that each type of chemical composition has correlation with equilibrium mineral assemblages, formed by metamorphism. He found that pelitic rocks contain aluminum rich mineral assemblages corderite, plagioclase, garnet and alumino-silicate polymorphs. Calcareous rock is found to contain diopside, wollastonite and Ca-amphibole.

In the Orijarvi region of southern Finland, Eskola noticed similar type of contact metamorphism, and found more or less similar kind of relationship between rock chemical composition and mineral assemblages. However, he found that the chemical composition of pelite is similar to Oslo region, but with different mineral assemblages. In Oslo region, the rock has K-feldspar and cordierite, whereas in Orijarvi region the rock has muscovite and biotite, suggesting different physical conditions of metamorphism in these regions. Eskola concluded that Finnish rocks equilibrated at low temperatures and Oslo rocks at high temperatures. Based on the occurrence of similar rock compositions and mineral assemblages worldwide, Eskola (1920/1921) proposed the concept of facies that is defined as:

> Metamorphic facies is to designate a group of rocks characterized by definite set of minerals, which under the conditions during their formation were in perfect equilibrium with each other. The qualitative mineral composition in the rocks of a given facies varies gradually in correspondence with variation in the chemical composition of the rocks.

In 1939, Eskola redefined the facies as "The significance of this principle is based on the observation that the mineral parageneses of metamorphic rocks in many cases confirm to the laws of chemical equilibrium, but the attainment of chemical equilibrium is not essential to the definition of metamorphic facies".

Tilley in 1924 comments that:

> A given facies may include rocks of widely different bulk composition, the variable of chemical composition is thus allowed for, whilst number of facies expresses the variable physical environment under which the rocks have been formed.
>
> This concept is a distinct advance over the classification adopted by Grubenmann, and one is impressed with the elasticity of the classification, for the number of so-called facies may be indefinitely increased as progress in the study of metamorphism may require.

The concept of facies is proposed based on the study mostly of mafic or basic rocks common in all orogeneic belts. Facies concept is purely descriptive in nature. Each facies covers wide range of P-T conditions and each facies have specific mineral assemblages for a specific rock chemical composition. The range of P-T condition suggested for each facies is purely interpretative. The P-T condition assigned in each facies is based on volume, density and volatile content of the minerals present. Eskola (1920) initially proposed five metamorphic facies, namely the greenschist facies, amphibolite facies, hornfels facies, sanidinite facies and eclogite facies. Subsequently, with more information available, he added three more facies the granulite facies, epidote-amphibolite facies, and glaucophane schist facies in 1939.

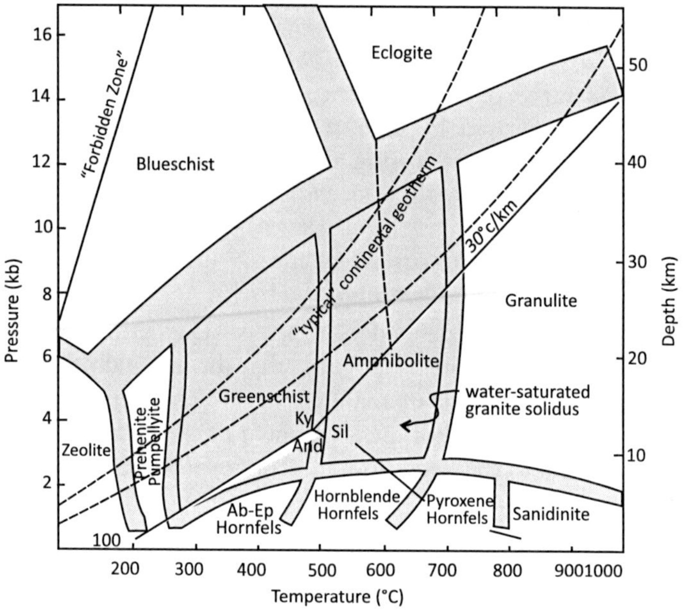

Fig. 8.1: Facies Diagram

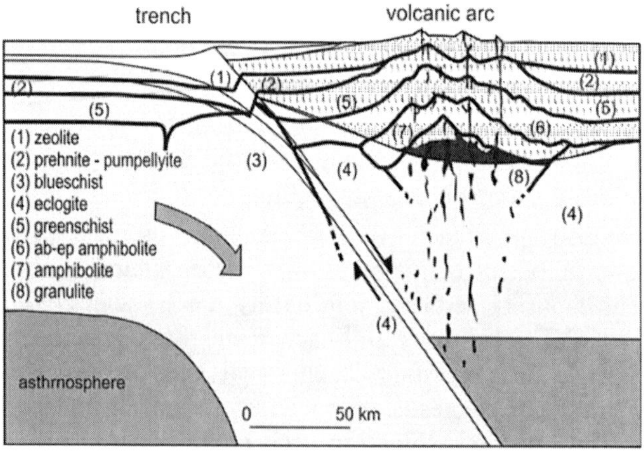

Fig. 8.2: Distribution of Facies in Subduction Zone

However, he changed the name of hornfels facies to pyroxene-hornfels facies. Coombs (1954, 1960, 1961) added two more facies, the zeolite facies and prehnite-pumpellyite facies after his study of burial metamorphism in New Zealand, Fyfe (1958) proposed two more facies, the albite-epidote hornfels facies and hornblende hornfels facies. The epidote-amphibole facies of Eskola is replaced by albite-epidote hornfels facies. Numerous other facies and sub-facies were suggested by many workers in the later years, but for simplicity, only eleven facies are finally adopted as given in Fig. 8.1. The distribution of facies in the subduction zone is shown in Fig. 8.2. It is possible to assign approximate P-T limits for each facies. The approximate P-T limits of each facies are given in Table 8.1. As mentioned earlier, the facies names are derived purely based on the minerals present in the basic rock. However, it is to be mentioned that the mineral assemblages found in low pressures are not significantly different from medium to high pressure assemblages in basic rock.

As an example, the albite-epidote hornfels facies mineral assemblage is same as that of greenschist facies assemblage. Similarly, hornblende hornfels mineral assemblage is same as that of amphibolite facies. Therefore, to retain the facies names, mineral assemblages from other rock chemical compositions are to be considered to define a particular facies. The P-T limit for each facies can be seen in Fig. 8.1. It may be noticed that each facies is separated from other facies by means of a small field, instead of a line, called the reaction boundary or isograd boundary. The isograd boundary may be a thin or thick field, depending on the bulk composition of the rock, presence of solid solution minerals involved in the reaction and composition of fluids. When crossing the boundary from one facies to another facies, a few new minerals appear or old minerals disappear. Table 8.1 shows pressure-temperature conditions estimated for regional and contact metamorphic rocks by thermobaromeric method using mineral assemblages mainly from peltic and basic rocks.

Table 8.1: Temperature-Pressure Range for each Facies in Regional and Contact Metamorphic Terrains

Name of Facies	Temperature Range (°C)	Pressure Range (kb)
Contact Metamorphic Facies		
Albite-epidote hornfels facies	400 – 450	0.0 to ~2.0
Hornblende hornfels facies	500 – 560 (Sometimes 550 – 700)	0.5 to ~2.0
Pyroxene hornfels facies	600 – 680 (Sometimes 650 – 800)	1.0 to ~4.0
Sanidinite facies	780 – 870 (Sometimes to 1000)	< 0 .5
Regional metamorphic facies		
Zeolite facies	200 – 400	~0.5 to ~4.5
Prehenite-pumpellyite facies	350 – 450	1.0 to ~6.0
Greenschist facies	500 – 550	~3.0 to ~8.0
Amphibolite facies	550 – 700	~3.0 to ~8.0
Granulite facies	700 – 800, upto 900	~3.0 to ~10.0
High Pressure facies		
Blue schist facies	200 – 400	~5.0 to ~10.0
Eclogite facies	550 – 900	~10.0

Yardley (1989) broadly divided the facies into four categories, namely facies of higher pressure, medium pressure, low pressure and very low pressure. These four types are described below.

8.1 High Pressure Facies

Blueschist and eclogite facies form the high pressure facies (Fig. 8.2). Blueschsit forms at low T and high P and is also known as glaucophane schist facies due to the presence of blue coloured sodic amphibole glaucophane/crosstie. Eclogite facies mineral assemblages are formed at medium T and very high P at the deeper part of the crust or in upper mantle part of the subduction zone. The rocks contain high pressure minerals jadeite and coesite. Below 1.5 GPa pressure and temperature of 600° to 800°C, granulite facies rocks are formed, and this is considered to represent medium pressure facies.

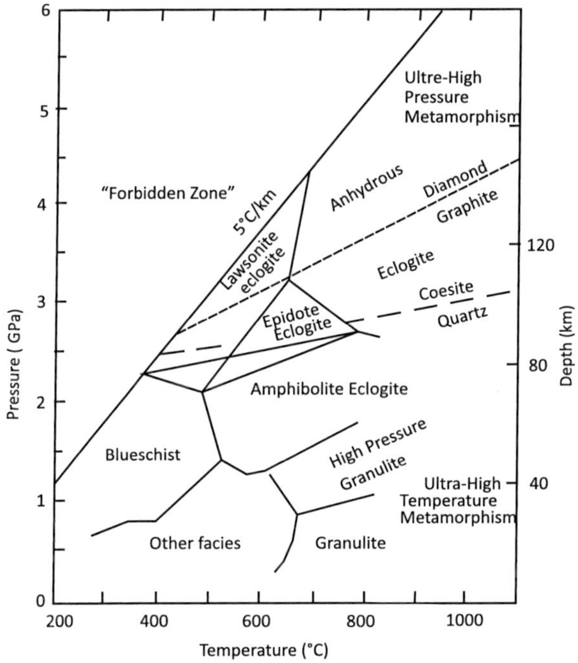

Fig. 8.3: P-T Diagram Showing High to Medium Pressure Facies

8.2 Medium Pressure Facies

Most metamorphic rocks found on the surface of the earth represent continental geothermal gradient that is responsible for the formation of greenschist, amphibolite and granulite facies assemblages. However, the granulite facies metamorphic rocks are found only in the eroded part of deeper continental cratons of pre-Cambrian age. Both basic and pelitic rocks show appropriate mineral assemblages in each facies. Greenschist facies of basic rocks contain actinolite, oligoclase plagioclase, epidote, chlorite, whereas in the same facies, pelites have garnet, biotite, muscovite, chlorite, quartz and oligoclase plagioclase. Similarly, amphibolite of basic rock has hornblende, andesine plagioclase and epidote, whereas pelites have garnet, biotite, staurolite, muscovite, cordierite, and andesine plagioclase and kyanite/sillimanite minerals. Granulite facies rocks have entirely different minerals with or without orthopyroxene in each of these compositions.

8.3 Low Pressure Facies

Albite-epidote hornfels, hornblende hornfels, pyroxene hornfels and sanidinite facies are product of contact metamorphism. These are developed at the intrusive contact of acidic plutons and country rock at shallow depths, producing contact metamorphic aureoles.

8.4 Facies of very Low Pressure and Temperature

Though zeolite and prehnite-pumpellite facies are recognized in the field, they fail to show equilibrium mineral assemblages. These facies are developed only in the regions where the geothermal gradient is fairly high with high fluid circulation. These facies are found only in volcanic terrains showing burial metamorphism and hydrothermal fluid activity. The development of these facies is more common in basic rock composition, as found in the volcanic rocks of New Zealand. These two facies show zeolite zone, Prehenite-pumpellyite zone, pumpellyite-actinolite zone, chlorite-clinozoisite zone. Albite at lower grade is replaced by a more calcic plagioclase oligoclase. The other rock compositions do not show any new minerals due to their improper chemical composition.

8.5 Mapping the Metamorphic Terrains

While mapping a metamorphic terrain, it is necessary to combine the facies with zones and isograds.

For example, if we come across greenschist facies assemblage in a metamorphic terrain, it is necessary to delineate chlorite zone, biotite zone and garnet zone for the same facies. Similarly, we have to define the zones and the reaction isograds, wherever possible, in each facies.

Reaction isograds may either be separating the zones within a single facies or may be separating one facies from another facies

For example, biotite zone may be separated from chlorite zone by a reaction isograd on the lower temperature side, and another reaction isograd may be separating biotite zone from garnet zone at higher temperature side.

Greenschist facies may be separated from amphibolite facies by a reaction isograd at medium temperature. Similarly, the amphibolite facies may be separated from granulite facies by another reaction isograd at still higher temperature.

By using a single type rock composition, it is not possible to define reaction isograds or zones or facies boundaries in the metamorphic terrain. If we have only basic rock in a metamorphic terrain, it is not possible to demarcate all the facies. We may be able to define only lower to middle temperature or low pressure to high pressure facies or zones. If we have more than one rock compositions occurring in a terrain, such as basic, pelites, and impure calcareous rocks, it may be possible to delineate number of zones and facies for such terrains in the map.

The development of facies in the metamorphic terrains is solely dependent on P-T condition experienced by that terrain. The P-T conditions experienced by one terrain is different from another terrain, because of their tectonic environment and geothermal gradient. In view of these, Miyashiro (1971) proposed the concept of facies series for prograde metamorphism. When we traverse along a metamorphic terrain with increase in temperature, one may come across a sequence of facies. The sequence of facies varies from one terrain to another terrain. Based on the sequence of facies variation in a given metamorphic terrain, he suggested five facies series, each named after the type locality. They include:

1. Contact facies series—very low pressure, and low to high temperature (< 1.0 kb; 200 to 800°C)
2. Buchan or Abukuma facies series—low pressure and low to medium temperature (~1.0 to 3.0 kb; 300 to 600°C)
3. Barrovian facies series—medium pressure and medium to high temperature (~2.0 to ~7.0 kb; 300 to 800°C)
4. Shanbagawa facies series—high pressure and low to medium temperature (~2.0 to 10.0 kb; 200 to 600°C)
5. Franciscan facies series—in California—(information not available)

Chapter 9

Phase Rule and its Application

The Gibbs phase rule is as follows: $F = C - P + 2$

F = degree of freedom

C = number components

P = number of phase

2 = intensive variables pressure and temperature

Before applying phase rule, one has to remember that the mineral assemblages are at equilibrium in a range of P-T condition. It means once the mineral assemblages are formed, they are at equilibrium and remain stable within the certain P-T range.

As per phase rule, in the common situation a rock at equilibrium, the number of phases should be equal to or less than number components. It is called Goldschmidt's mineralogical phase rule. Consider the following examples:

Example 1: number of components (C) = 3, Number phases (P) = 3, Variables = 2,

The degree of freedom becomes to (F) = 2, it means that three-phase assemblage will be at thermodynamic equilibrium in a range of temperature and pressure. It is called divarient or bivariant assemblage.

Example 2: number of component (C) = 3, Number phases (P) = 4, variables = 2

The degree of freedom becomes (F) = 1. In this case, all four phases are dependent on pressure and temperature variation. It means that four phases occur along a line where the temperature is dependent on pressure or vice-versa. So this becomes an univarient assemblage and coexist only along a line in the P-T diagram.

Example 3: number of component (C) = 3, number of phases (P) = 5, Variables = 2 degree of freedom becomes (F) = 0

Here all five phases coexist at a point in the P-T diagram. All phases are at equilibrium only at a fixed temperature and pressure. This becomes an invariant assemblage.

9.1 Compositional or Chemographic Diagram

In the rock, there may be many minerals of different chemical composition. The total number of elements or molecules covering all the minerals may be around 10 to 12. The molecules include MgO, FeO, CaO, MnO, K_2O, Na_2O, Al_2O_3, SiO_2 and H_2O. There may be other minor molecules like P_2O_5, TiO_2, which are not important to consider in metamorphic rocks. If a mineral is to be plotted in the chemographic diagram, the minerals with more than four components cannot be represented in two-dimensional space. A maximum of three components can be plotted in two dimensional space. It means that the mineral with three or two components can be plotted either within the triangular space or along a line. If you want to represent the fourth component in the two dimensional space, it is possible to represent in the tetrahedral space. In the tetrahedron, one of the point is taken as projection point. From this point, the minerals with the fourth component are projected and plotted within the triangular space. Beyond four components, if the rock system contains n number of components, it is theoretically possible to represent in n-dimensional space using mathematical approach. Therefore, in a rock system, though the number of components may be more than 4, the components are to be reduced by some way either to three or four components, so that the minerals of the rock can be plotted in the triangular or tetrahedral diagrams.

To reduce the number of components to 3, the following principles of Goldschmidt are to be followed,

- Fe^{2+}, Mg^{2+} and Mn^{2+} are divalent cations having similar ionic radius and charge (+2).

Similarly, K^+, Na^+ and Ca are having more or less similar ionic radius and monovalent ions, except Ca^{2+}. As these elements are having same charge and ionic radius, they can easily replace each other. Therefore, these components are combined and taken together as one component.

- Most of the metamorphic rocks commonly contain quartz. If the rock has silicate minerals and quartz, it indicates that the rock is saturated in SiO_2. Therefore, SiO_2 is eliminated while plotting the silicate minerals in the diagram.
- The volatiles, being a fluid, are not represented in the diagram, except in a few special cases.

Based on these principles, three compositional or chemogarphic diagrams are proposed by various workers to understand equilibrium mineral assemblages in each metamorphic facies.

9.2 Chemographic Plotting

For example, two components X and Y are plotted in a line, where left and right ends are X and Y. The phases C and D occurring in between X-Y are shown within this line (Fig. 9.1). Four reactions are possible, each reaction with three phases, and are as follows:

X + Y = C	50 X : 50 Y ratio for C	absent phase (D)
C + Y = D	50 C : 50 Y ratio for D	absent phase (X)
X + D = C	66 D : 34 X ratio for C	absent phase (Y)
X + Y = D	25 X : 75 Y ratio for D	absent phase (C)

Fig. 9.1: Binary Line having Two Components X and Y

If all the four phases are present, it can be plotted only as a point in the P-T diagram according to phase rule

$C + 2 = P + F,$

$F = 4 - 2 + 2 = 0.$

The degree of freedom is 0. Therefore, all the four phases can coexist at a point, that is an invariant point. In other words, all four phases can coexist only at a spectfic P-T. If

P-T is changed, three phases will coexist along a line called univariant line. Likewise we get four univariant lines. All these lines originate from the invariant point. If either temperature or pressure is changed independently, two coexist in a space called divariant field. We get four divaiant fields with two phases in each field. The orientation of the four univariant lines and the corresponding divariant fields cannot be done randomly in the P-T diagrams. Because each univariant line and the corresponding phases/minerals involved in the univariant line occur only along a specific P-T range. Likewise, the divariant mineral assemblages occur in certain P-T space. To plot all four univariant lines and divariant assemblages, specific geometrical approach developed by Schreinemakers is to be followed, and it is known as Schreinemakers' principle. It is as follows:

1. Take any two univariant lines X + Y = C (D), C + Y = D (X) in the P-T space as shown in Fig. 9.2a. The absent phases D and X are given in paranthesis.

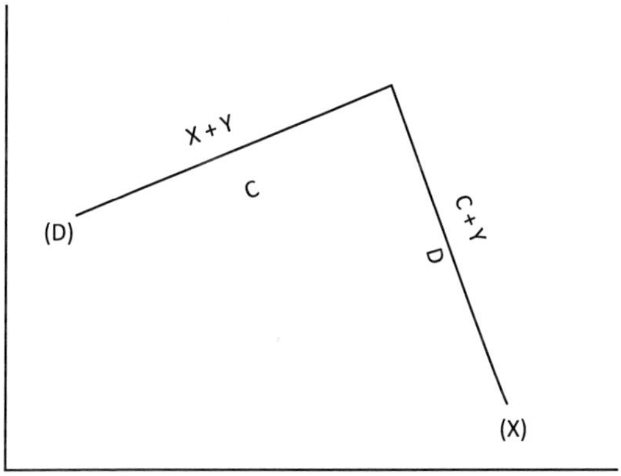

(a)

Fig. 9.2 (a): Schreinemakers' Diagram

In (D) absent line X + Y is located on the upper side of the line and C on the lower side of the line. Absent phase is given at the end of this line in bracket as (D). As per rule, it should be inferred that univariant lines with (X) and (Y) absent phases should orient on the bottom or southern side of (D) absent line, and the (C) absent univariant line on the top or northern side of (D) absent line (Fig. 9.2b). On the right side of (X) line, the line

(Y) should be oriented. As mentioned earlier, (C) absent line is oriented on the top side of (D) absent line.

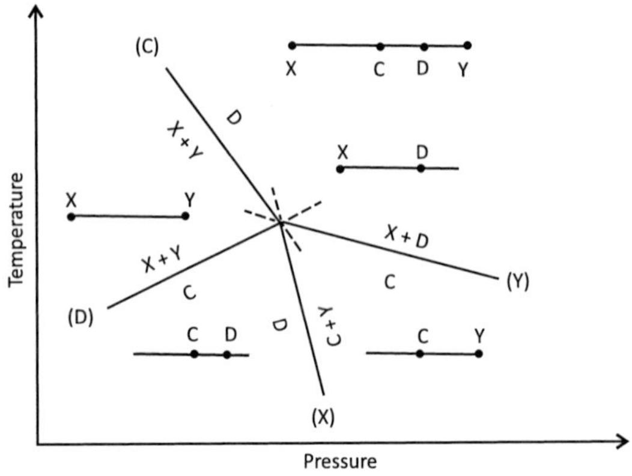

Fig. 9.2 (b): Schreinemakers Diagram for Two Component System

2. The univariant lines and their corresponding reactions are derived based on petrographic study done for equilibrium mineral assemblages in diffenet metamorphic zones. However, the slope of the univariant lines in the P-T space is drawn based on thermodynamic data calculated for the minerals (dP/dT slope) that participate in the univariant reactions. Field and petrographic data of mineral assemblages from various metamorphic grades are useful in the orientation of slope of the reactions. The space between two adjacent univariant lines indicates equilibrium mineral assemblages in divariant surface.
3. Five minerals A, B, C, D and E are plotted in the triangular diagram XYZ (Fig. 9.3 a).

   These minerals coexist at an invariant point. Five univariant reactions are possible with these five minerals.
4. Reaction No. 5 (C+B = E+D (A)) univariant line is plotted in the imaginary P-T space, where (E) and (D) absent univariant lines lie on the upper side, and (C) (B) absent univariant lines lie on the lower side, as shown in Fig. 9.3 (b). In (A) absent univariant line, C+B present on lower side of the line and E+D on the upper side of the line. To plot the next line in the P-T space, (C) and (B) absent univariant lines should be drawn on the lower

side of (A) absent line, and (E), (D) absent univariant lines on the upper side of (A) absent line. This is shown in Fig. 9.3 (c).

5. Therefore, for any univariant line (X), the phases occur on its upper side should be the absent phase on the lower side of reaction (X). Similarly, the phases that occur at the lower side of reaction (X) should be absent phase on the upper side of reaction (X).
6. This method of construction is followed for the above XYZ three-component system with five phases. Such construction is shown in Fig. 9.3 (d).
7. All the five univariant reactions are given here

(1) B+C = A+D (E); (2) E = A+C+D (B); (3) E = A+B+C (D); (4) A+D= E+B (C); (5) C+B = E+D (A)

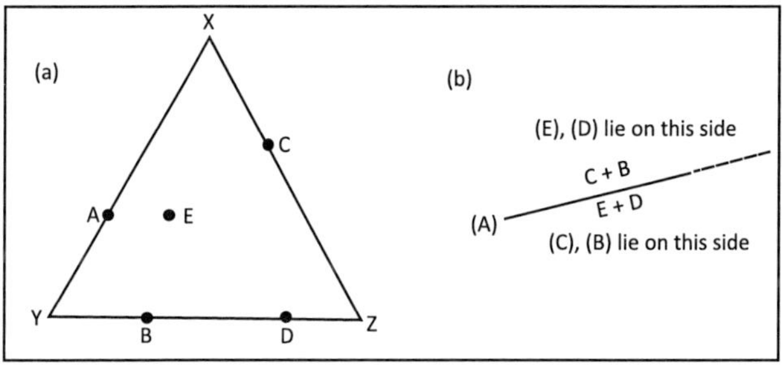

Fig. 9.3 (a) and (b)

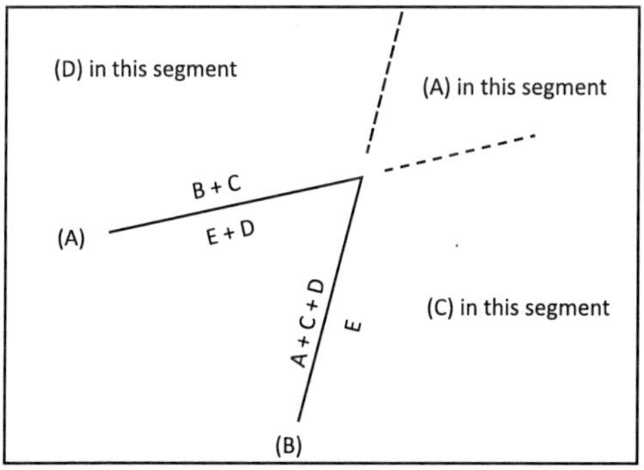

Fig. 9.3 (c)

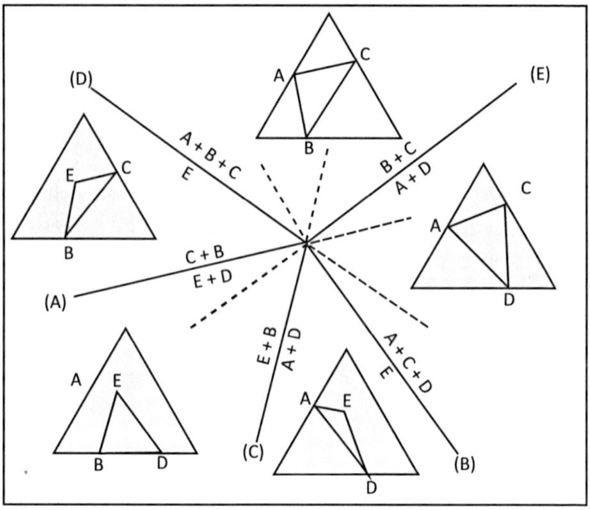

Fig. 9.3 (d): All five Univariant Lines are Oriented as per Schreinemakers' Rule and Shown

Each divariant surface is shown by equilibrium mineral assemblage in XYZ triangle.

For example, three-component system with CaO, Al_2O_3, SiO_2 is chosen for mineral plots and to show equilibrium mineral assemblages (Fig. 9.4). The minerals with these components are plotted in the triangle depending on their ratio of each mineral. The rock has the following minerals:

Wolllastonite – CaO . SiO_2

Anorthite – CaO . Al_2O_3 . $2SiO_2$

Sillimanite – Al_2O_3 . SiO_2

Quartz – SiO_2

Corundum – Al_2O_3

All the minerals have either one component, or more than one component CaO, Al_2O_3 and SiO_2. Each corner of the triangle is represented by CaO, Al_2O_3 and SiO_2 (Fig. 9.4). One corner is represented by 100% CaO likewise the other corners with Al_2O_3 and SiO_2. When you go from 100% CaO corner to other corners of the triangle, CaO reduces gradually and becomes zero at SiO_2 or Al_2O_3 corners. The same thing is true for the other two components.

Minerals corundum and quartz have 100% Al_2O_3, and 100% SiO_2 and plot at Al_2O_3, and SiO_2 corners. Wollastonite has got 50% CaO and 50% SiO_2, sillimanite - 50% Al_2O_3 and 50% SiO_2. Therefore, these two minerals plot at the mid-points between CaO-SiO_2 and Al_2O_3-SiO_2 lines. Minerals grossularite and anorthite have 3CaO.1 Al_2O_3.3 SiO_2 (43:14:43 ratio) and 1CaO : 1Al_2O_3 : 2SiO_2 (25:25:50 ratio) respectively, and plot within the triangle (Fig. 9.4).

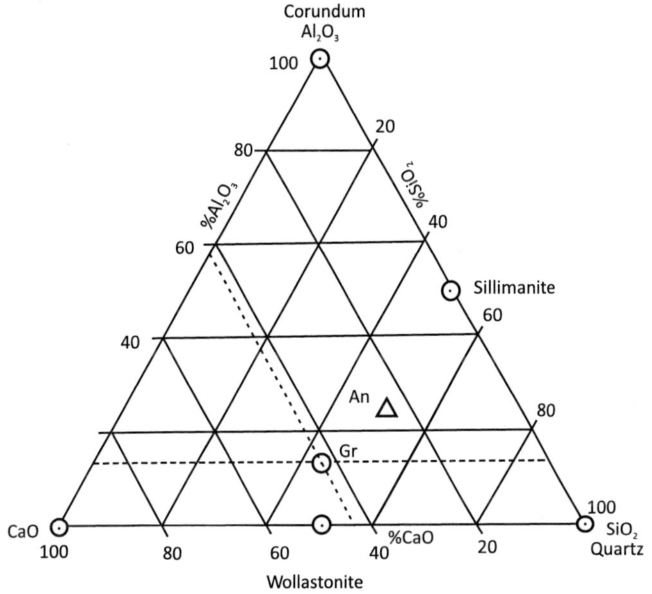

Fig. 9.4: Triangular Plot for CaO-Al_2O_3-SiO_2 Components

Now the equilibrium mineral assemblages are as follows and this is valid only for a particular metamorphic grade or zone (Fig. 9.5 a).

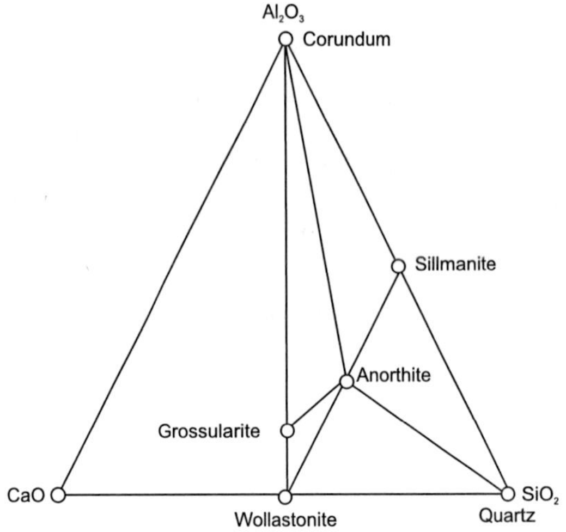

Fig. 9.5 (a): The Equilibrium Mineral Assemblages for a Particular Metamorphic Grade

Corundum – grossularite – anorthite;
Corundum – anorthite – sillimanite;
Anorthite – sillimanite – quartz;
wollastonite – anorthite – quartz;
Grossularite – wollastonite – anorthite

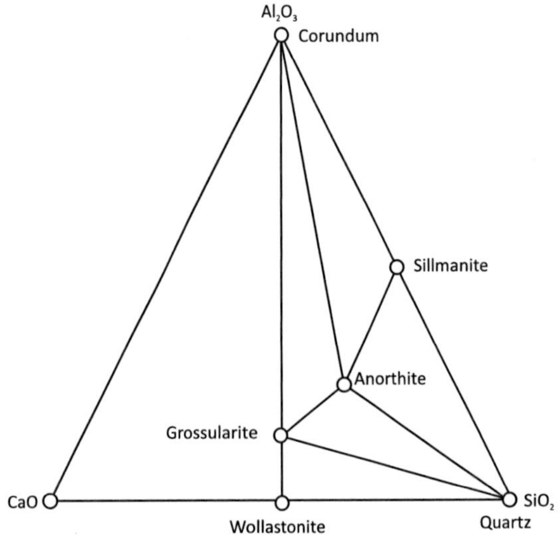

Fig. 9.5 (b): Equilibrium Mineral Assemblages for Another Metamorphic Grade

In this metamorphic zone, anorthite and wollastonite occur with quartz or grossularite but not with sillimanite. When we go from this zone to another zone, the mineral parageneses observed earlier is changed to the following assemblages (Fig. 9.5b):

Corundum – grossularite – anorthite;
Corundum – anorthite – sillimanite;
Grossularite – anorthite – quartz;
Grossularite – wollastonite – quartz;
Sillimanite – anorthite – quartz
Such change is explained by a reaction
anorthite + wollastonite = Grossularite + quartz

In Figs. 9.5(a) and 9.5(b), it can be observed that the tie line between wollastonite and anorthite (Fig. 9.5a) is broken, establishing a new tie line between grossularite and quartz (Fig. 9.5b). Figures 9.5(a) and 9.5(b) show changes in mineral parageneses and indicate two different metamorphic grades. The mineral assemblages are continuously changed from one grade/zone to another grade/zone, and are well observed in prograde metamorphic terrains. For a particular metamorphic zone, the paragenesis is distinct and equilibrium mineral assemblages are shown by connecting tie lines in the respective compositional diagram.

It may be pointed out that in the above-mentioned three-component system, six phases are present. As per phase rule, ($P + F = C + 2$), theoretically, for three components, only five phases are to be present to get an invariant point. Since the system has six phases, it should be having more than one invariant point. If five phases form an invariant point, there are six five-phase assemblages and hence there will be six invariant points. System with more than C+2 phases are known as multisystems, with more invariant points, univariant lines and divariant surfaces.

Each invariant point has five univariant lines and five divariant surfaces. Practically, for six invariant points, there will be 30 univariant lines and 30 divariant surfaces. However, many of the univariant lines and divariant surfaces are common to one or more invariant points, therefore, only a few univariant lines and divariant surfaces are possible. Some of the invariant points and univariant lines may become metastable for an extraordinary P-T situation. To understand such multisystems with a system of components with multiple reactions, suitable P-T grids are developed by researchers which is commonly termed as petrogentic grid. Several grids are developed for different rock compositions using computer program. The petrogenetic grids are developed for various parameters such as P-T, f_{O_2}-T and T-X, etc., using simple geometric rules of Schreinemakers. If the system has more than three components (four or five or n number of components), several reactions occur and constructing all the reaction lines and invariant points are not possible in a phase diagram. Therefore, it is done by computer programming, developed by Zen (1964), Day (1984), Spear (1990) and others, using linear algebra and topology. For details, read relavant books and research papers of these authors.

9.3 Tie Line

In the compositional diagram, if the minerals are connected with one another by means of a line, it is called as tie line. This means that the minerals connected by tie lines represent equilibrium assemblages for a particular metamorphic zone. Sometimes, mineral or minerals may be shown in the form of a field instead of single point, as shown in Fig. 9.6 for biotite, almandine and staurolite. It is seen that composition of biotite has large variation; staurolite has less variation and almandine with too little variation. It means that each mineral has compositional variations in three or four components of the respective system. If two minerals have compositional variations in their respective components, they are connected by several tie lines. In Fig. 9.6, biotite and almandine are connected by many tie lines. One specific biotite composition is connected by a line with one specific garnet composition. Second biotite composition point is connected with second garnet composition point. Likewise, other lines are connected between these two minerals. This is valid for all other tie lines connected between two minerals in Fig. 9.6. In other words, a pair of minerals coexists only with specific mineral compositions, as shown by a particular tie line in any metamorphic zone. This type of compositional variation in minerals is due to variation in bulk rock chemical composition.

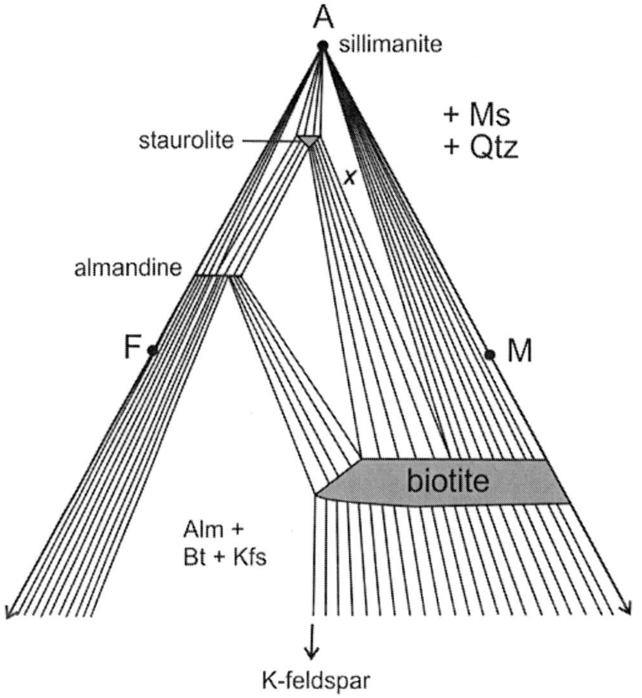

Fig. 9.6: AFM Diagram Showing Sill-Bt-St; Alm-Bt-St Assemblages of the pelites. St-Bt, Alm-Bt, Bt-Alm are connected by many tie lines. In St-Bt tie line, each line is connected by specific St and Bt compositions. Likewise, for all the mineral pairs shown here

9.4 Solid Solutions

Most of the minerals are not pure end members but are solid solutions. For example, garnet has a composition of three (Mg, Fe, Mn, Ca)O.Al_2O_3.3SiO_2. It is a mixture of four end members such as pyrope (Mg member) almandine (Fe member), spessartine (Mn member) and grossularite (Ca member). The end-members replace one another in solid state through grain boundary diffusion depending on the physical conditions of metamorphism. As exchange of end member components is taking place in solid state, it is known as solid solution.

9.5 Chemographic Diagrams and their Application

As seen earlier, the composition of parent rocks and their mineralogy are important to construct different types of chemographic diagrams. The most common sedimentary rocks that have undergone metamorphism include pelites, greywackes and altered basic-ultrabasic rocks. Based on the mineral parageneses of different metamorphic rocks, three basic diagrams, viz. ACF, AKF, AKFM (AFM) are proposed. These diagrams show, mineral parageneses, mineralogical changes in each metamorphic grade or zones and the possible metamorphic reactions that brought those changes.

Example:

A metamorphic rock contains the following mineral assemblage:

Andalusite	–	$Al_2 SiO_5$
Anorthite	–	$CaAl_2 Si_2O_8$
Cordierite	–	$2 (Mg, Fe)O{\cdot}2Al_2O_3{\cdot} 5SiO_2$
Grossularite	-	$Ca_3 Al_2 (SiO_4)_3$
Diopside	–	$Ca Mg Si_2O_6$
Hypersthene	–	$(Mg, Fe)_2 Si_2 O_6$

Looking into these minerals, one may see that minerals contain five components—CaO, MgO, FeO, Al_2O_3 and SiO_2. To arrive at three components and to plot these minerals in the triangular diagram, Goldschmidt's rules are to be followed. All the minerals have SiO_2, and so it is eliminated. Among the remaining four components, most minerals have Al_2O_3 and CaO, and therefore, these two components are taken independently. As FeO and MgO can replace each other, as per the Goldschmidt's rule, MgO and FeO are taken together as FeO component. As MgO is relatively less than FeO in these minerals, FeO component is taken instead of MgO. Three components Al_2O_3, CaO and FeO are taken to plot these minerals in the diagram. The components are abbreviated as

A for Al_2O_3,

C for CaO,

F for FeO.

This diagram is known as ACF diagram. This was proposed by Eskola in 1939. Sometimes, the above-mentioned minerals may have some MnO, Fe_2O_3, and TiO_2. Since TiO_2 replaces SiO_2, Al_2O_3 replaces Fe_2O_3, MnO replaces FeO, all these components are either appropriately included or eliminated while plotting the minerals.

9.6 ACF Diagram

This diagram is most useful for the rocks having minerals rich in calcium, aluminum and iron; this diagram is suitable for metamorphic marls and mafic rocks. The minerals that plot in ACF diagram is shown in Fig. 9.7.

If the rock contains minerals like muscovite, K-feldspar in appreciable amounts this diagram is not suitable.

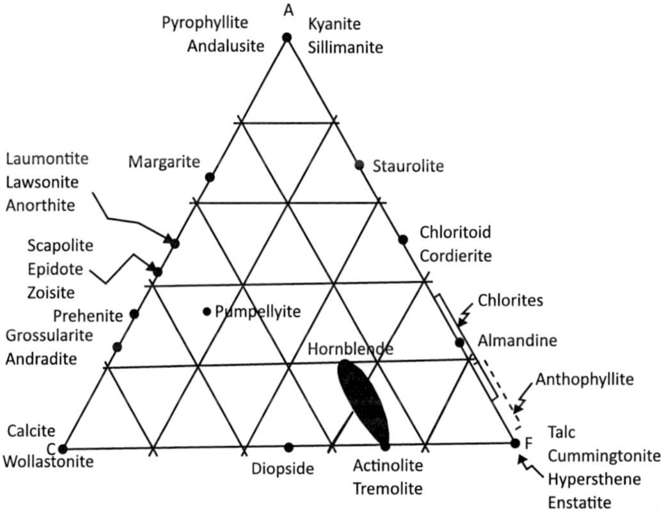

Fig. 9.7: Minerals in ACF Diagram

The following minerals are plotted in ACF diagram:

Mineral	Composition	A C F percentage
Very rich in Al		
Pyrophyllite	$Al_2[(OH)_2/Si_4O_{10}]$	A = 100
Andalusite, kyanite, sillimanite	Al_2SiO_5	A = 100

Mineral	Composition	A C F percentage
Rich in Al and Mg, Fe		
Staurolite	4 $FeO.9 Al_2O_3.8 SiO_2. H_2O$	A = 69, F = 31
Cordierite	2 $(Mg, Fe)O_2Al_2O_3. 5SiO_2$	A = 50, F = 50
Chloritoid	$FeO. Al_2O_3.SiO_2.H_2O$	A = 50, F = 50
Rich in Al and Ca		
Margarite	$Ca Al_2[(OH)_2/Si_2Al_2O_{10}]$	A = 67, C = 33
Laumonite	$Ca [Al_2Si_4O_{12}]4H_2O$	A = 50, C = 50
Lawsonite	$Ca Al_2 [(OH)_2/Si_2O_7]H_2O$	A = 50, C = 50
Anorthite	$Ca Al_2Si_2O_8$	A = 50, C = 50
Epidote group	4 $CaO.3 (Al, Fe^{3+})_2O_3.6 SiO_2.H_2O$	A = 43, C = 57
Pumpellyite	Similar to epidote, but contains Mg and Fe^{2+}. Approximate composition is $8CaO_2$ (Mg, Fe, Mn)O_5 Al, Fe)$_2O_3$.12 SiO_2.7H_2O	A = 34, C = 53, F = 13
Prehenite	$Ca_2 [(OH)_2/Al_2Si_3O_{10}$	A = 33, C = 67
Rich in Ca		
Grossularite – andradite	$Ca_3 Al_2 (SiO_4)_3 - Ca_3 Fe_2(SiO_4)_3$ (note that A comprises Fe^{3+} substituting for Al)	C = 75, A = 25
Wollastonite	$CaSiO_3$	C = 100
Calcite	$CaCO_3$	C = 100
Rich in Mg and Fe^{2+}		
Diopside-hedenbergite	$Ca MgSi_2O_6 - Ca FeSi_2O_6$	F = 50, C = 50
Dolomite	$Ca Mg(CO_3)$	F = 50, C = 50
Tremolite	Ca_2 (Mg, Fe)$_5$[(OH)$_2$/Si_8O_{22}]. Only up to 20 mole% of Mgo is replaced by FeO	F = 71.5, C = 28.5
Actinolite	Similar to tremolite, but containing more Fe, and a little Mg, + Si is replaced by 2 Al	
Anthophyllite-gedrite	Similar in composition as Tremolite-actinolite, but orthorhombic instead of monoclinic	F = 100

Mineral	Composition	A C F percentage
Enstatite-hypersthene	$Mg_2Si_2O_6$ – $(Mg, Fe)_2Si_2O_6$	F = 100
Talc	Mg_3 $[(OH)_2/Si_4O_{10}]$	F = 100
Almandine	Fe_3 Al_2 $(SiO_4)_3$	F = 75, A = 25
Pyrope	Mg_3 Al_2 $(SiO_4)_3$	F = 75, A = 25
Chlorite	Varibale composition Mg_5 (Mg, Al)$[(OH)_8/(Al, Si)$ $Si_3O_{10}]$ upto (Mg, Fe)$_4$ $Al_2[(OH)_8/Al_2$ $Si_2O_{10}]$	F = 90-65, A = 10-35
Minerals containing alkali that cannot be represented in an ACF diagram		
Glaucophane, crosstie Na-amphibole	Na_2 (Mg, Fe)$_3$(Al, Fe)$_2[(OH)_2/Si_8O_{22}]$	
Jadeite pyroxene	$NaAlSi_2O_6$	
Muscovite	K Al_2 $[(OH)_2/Si_3AlO_{10}]$	
Phengite	Similar to muscovite but has more Si + Mg and less Al	
Paragonite	Na Al_2 $[(OH)_2/Si_3AlO_{10}]$ also as solid solution with muscovite	
Biotite	K (Mg, Fe, Mn)$_3[(OH)_2/$ $Si_3AlO_{10}]$	
Phlogopite	K Mg_3 $[(OH)_2/Si_3AlO_{10}]$	
Annite	K Fe_3 $[(OH)_2/Si_3AlO_{10}]$	

To plot rocks rich in calcium, aluminum, the following steps and corrections are to be made.

1. Whole rock chemical analysis for all major oxides CaO, MgO, FeO, MnO, Na_2O, K_2O, Al_2O_3, SiO_2, P_2O_5, TiO_2. These 10 oxides are to be reduced to three components
2. Model analysis for the presence of various accessory minerals such as magnetite, hematite, sphene, apatite, etc. from petrographic study of thin sections for the chemically analyzed rock.
3. From the weight (wt%) analysis of the oxides and as per the model % of various accessory/minor minerals as in point 2, the oxides data are to be recalculated by the following method:

Magnetite: $FeO \cdot Fe_2O_3$ = 30 wt. % FeO and 70 wt. % Fe_2O_3

Whole rock — total FeO – 3.2%; total Fe_2O_3 – 0.4% determined from rock ananlysis.

Modal magnetite in the rock is 1.5%.

From this whole rock value, magnetite modal data is used to determine FeO: Fe_2O_3 (3:7) ratio.

The amount to be deducted from total FeO and total Fe_2O_3 for magnetite is

FeO = [3.2 – ((1.5 × 3.20/100) = 0.048%)] 3.2 – 0.048 = 3.152%

Fe_2O_3 = [0.4 – ((1.5 × 0.4/100) = 0.006%)]

0.4 – 0.006 = 0.394%

In this manner, the corrections are to be done for ilmenite and sphene

Ilmenite: $FeO \cdot TiO_2$ = 50 wt. % FeO × modal % of ilemnite – to be deducted

Sphene: CaO. TiO_2. SiO_2 = 30 wt. % CaO × modal % of sphene – to be deducted

(Ideal composition contains 35% CaO, but sphene contains less CaO)

4. After all the corrections are made from wt% oxides, the remaining %age of oxides are added together and recalculated to 100% for all the three components and plot the rock compositions.
5. Once moles are calculated, the following further corrections are to be made to arrive at A, C and F data

   $[Al_2O_3] + [Fe_2O_3] - ([Na_2O] + [K_2O]) = A$

   $[CaO] - 3.3\ [P_2O_5] = C$

   $MgO] + [MnO] + [FeO] = F$

If minerals are to be plotted in ACF diagram, the composition of individual minerals is analyzed. Correction and re-calculation to 100 is done for three components (A, C, F) and then plotted.

Example: Garnet composition – $(Mg_{1.4}, Fe_{1.9}, Mn_{0.1}, Ca_{0.6})_3$ Al_2 Si_3 O_{12}

Rewritten as	MgO	1.4
	FeO	1.9
	MnO	0.1
Total	CaO	0.6 (Total 3.0)
	Al_2O_3	1.0
	SiO_2	3.0
	Ratio 3 : 1 : 3.	

However, to plot this mineral, SiO_2 is not considered. It is assumed that the rock is saturated with silica. The following recalculation is done.

F –	(MgO + FeO + MnO) = (1.4 + 1.9 + 0.1) = 2.4
C –	CaO 0.6
A –	Al_2O_3 1.0
Total	4.0

Recalculated to 100% F = 60%; C = 15%; A = 25%

Similar exercise is to be done for other minerals, before plotting in ACF or AKF or AFM or any such diagrams.

9.7 AKF Diagram

This diagram is to plot minerals with significant amount of K and less amount of Ca and Mg. In this diagram, C is replaced by K, and therefore known as AKF diagram. The minerals with K such as K-feldspar, muscovite, biotite, stilpnomelane, etc. can be plotted. However, Ca bearing minerals cannot be plotted in such diagram. If the rock has minerals with Al, K, Ca and Fe components, both ACF and AKF diagrams are to be used simultaneously.

The same method of correction, as mentioned in ACF diagram, is to be followed, for plotting rock composition and minerals in AKF diagram. After necessary correction, recalculation is to be done to get A : K : F ratio, as per the following formula:

$A = [Al_2O_3] + [Fe_2O_3] - ([Na_2O] + [K_2O] + [CaO]$

$K = [K_2O]$

$F = [FeO] + [MgO] + [MnO]$

$A + K + F = 100$

Correction for A for the following minerals from model analysis:

1/3 of the [CaO] in grossularite, andradite;

¾ of the [CaO] in the zoisite, epidote

½ of the [CaO] in anorthite

Twice the [CaO] in margarite.

$A = [Al_2O_3] + [Fe_2O] - ([Na_2O] + [K_2O]$

$K = K_2O$

$F = [FeO] + [MgO] + [MnO]$ – (correction with regard to hornblende and diopside)

$A + K + F = 100.$

If the rock contain ACF and AKF minerals, then both the diagrams are to be used simultaneously (Fig. 9.8). Even using these diagrams, the pelitic rock cannot be represented due to large variation in Mg and Fe, both in whole rock composition as well as in most of the minerals.

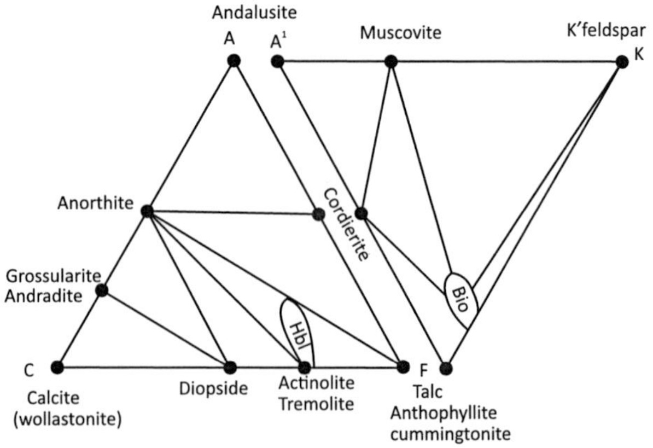

Fig. 9.8: ACF and A′KF Diagrams—for the Rock with Ca- and K-Bearing Minerals

9.8 AFM Diagram

In ACF and AKF diagrams, FeO, MgO, MnO are grouped together as one component F. This becomes difficult to represent the minerals chlorite, biotite, muscovite, garnet, staurolite, cordierite, andalusite, etc. due to over simplification and treating MgO and FeO as one component.

Though Mg and Fe substitute with each other, the extent of substitution entirely depends on the bulk rock composition, pressure and temperature of metamorphism. In many cases, the substitution between Mg and Fe is limited. Therefore, Mg and Fe component should be treated separately along with Al and K. In that case, it becomes a four-component system. In this system, Ca-bearing minerals cannot be plotted. Moreover, it is a four-component system, and therefore, not possible to plot the minerals in two-dimensional space. These four components can be represented only in a tetrahedron. To show tetrahedron in two-dimensional space, one component of the tetrahedron is to be taken as a projection point. In this case, K is taken as the projection point, because pelites have either plenty of muscovite or K-feldspar. Therefore, for plotting minerals in low to medium grade rocks, muscovite is used as projection point and for higher grades k-feldspar as projection point. From this point, the minerals with all four components, plot within the tetrahedron, are projected. In such projection, minerals with four components fall outside AFM triangular space of the tetrahedron (Fig. 9.9).

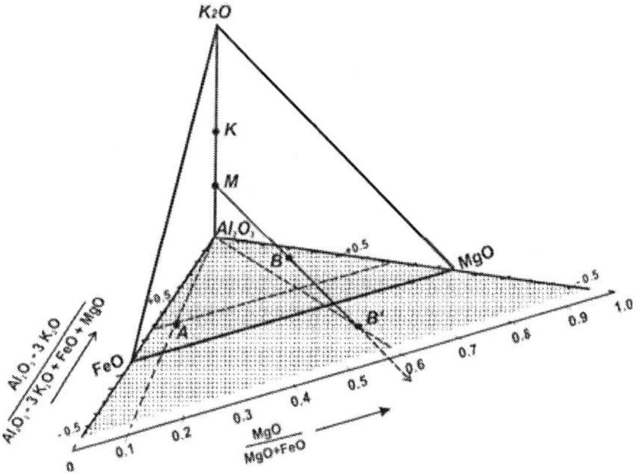

Fig. 9.9: AFM Tetrahedral Plot Projected from Muscovite Point (M)

Pelite is the most common metamorphic rock that contains muscovite, K-feldspar, sillimanite, kyanite, cordierite, staurolite, plagioclase, biotite, garnet and other minerals. Muscovite and K-feldspar have K and Al compositions. Therefore, one of the K-bearing minerals is used as a projection point, and other minerals with the remaining three components are plotted accordingly. Basic rocks cannot be plotted in AFM diagram due to the presence of large amount of Ca-bearing minerals.

In the AFM diagram calculation, two points are to be taken into account to derive A. If the rock contains more muscovite, then muscovite should be used for getting A. If the rock is devoid of muscovite but K-feldspar, then K-feldspar should be used for getting A. The following calculation is done from the chemical data of the minerals to plot in the AFM diagram.

In case of muscovite as projection point, A should be calculated by:

$$A = \frac{Al_2O_3 - 3K_2O}{Al_2O_3 - 3K_2O + MgO + FeO}$$

$$M = \frac{MgO}{MgO + FeO}$$

In case of K-feldspar as projection point, it is done by

$$M = \frac{Al_2O_3 - K_2O}{Al_2O_3 - K_2O + FeO + MgO}$$

To derive FeO, from the chemical data of the minerals, Fe_2O_3, and TiO_2 moles are to be deducted from total FeO. Then M data is calculated

$A' = Al_2O_3 - (K_2O + Na_2O + CaO)$

$F = FeO - Fe_2O_3\ (-\ TiO_2)$

$M = MgO$

Chapter 10

Metamorphism of Basic or Mafic Rocks

10.1 Introduction

Before going into the details of metamorphism for various rock compositions, a few basic concepts are to be understood. During prograde metamorphism, some minerals may appear or disappear. In earlier times, appearance of new mineral in a particular grade or zone is termed as index mineral. However, it is not used nowadays because the mineral or minerals, used as index minerals, need not necessarily form only at the beginning of a metamorphic zone or grade. They may form either at lower P-T or at higher P-T of a particular zone and are entirely dependent on the bulk rock composition and the kind of fluids present in the rock. This may be well understood in the next few sections, in which how the rock composition plays a significant role in the formation of new minerals. Therefore, the index mineral concept is not taken into consideration in the forthcoming chapters. Only three rock compositions, viz. basic, pelitic and calcareous are discussed in little detail for the better understanding of contact and regional metamorphism.

Volcanic rocks of basaltic to andesitic compositions, gabbro, diorite and immature mafic to greywackes may have been affected by regional metamorphism. Among these, the most common ones are the volcanic rocks occurring in a variety of tectonic settings like the oceanic crust, volcanic arcs and rift zones, etc.

The mafic rocks contain usually the anhydrous minerals pyroxene, plagioclase and olivine in variable proportions.

If such rocks have to undergo metamorphism, hydration is a pre-requisite for metamorphism. If water is not available, the mafic rocks are not easily affected by metamorphism. Therefore, before the initiation of metamorphism, the minerals in mafic rock have to be hydrothermally altered to low temperature minerals such as zeolites, chlorites, etc. If adequate water is available, the alteration may be completed, provided the kinetic factors do not retard the alteration. The mafic rock minerals are chemically more complex with extensive solid solutions. Most of the reactions are continuous reactions. Depending on the type of weathering and composition of circulating fluids, pyroxene and plagioclase are converted to low temperature minerals such as low temperature amphibole, epidote, chlorite, zeolites, etc. Labradorite (An_{50-70}; Ca plagioclase) is converted to albite with minimum anorthite content (An_{0-7}). Calcium released from plagioclase produce minerals such as calcite, epidote, titanite and low temperature amphibole. This factor led to the formation of only a few new minerals in this rock. Appearance and disappearance of minerals in prograde condition may vary depending on the bulk composition of the rock. However, it is different in pelites that show more number of new minerals during prograde metamorphism.

10.2 Contact Metamorphism

The weathered basic country rock is affected by acidic igneous intrusion and produce contact metamorphism. The mineral assemblages found in contact zones are very much similar to low to medium temperature and low pressure metamorphism of regional metamorphism in basic rocks. The mineral assemblages in albite-epidote facies of contact metamorphism are similar to the assemblages of zeolite facies and prehenite-pumpellyite facies of regional metamorphism. Likewise, hornblende hornfels facies assemblages are similar to greenschist facies, and pyroxene hornfels facies assemblages to that of amphibolite facies. The development of all regional metamorphic facies mineral assemblages is discussed in detail in the forthcoming pages. Therefore, one can refer to regional metamorphic facies description to understand contact metamorphic facies of basic rocks.

However, it is not the same in the contact metamorphism of pelites. The mineral assemblages in each facies of contact and regional metamorphic facies are quite different and dealt in detail in the pelites chapter.

10.3 Regional Metamorphism

Regional metamorphism of mafic rocks is considered broadly in two types, namely (a) Barrovian metamorphic series and (b) high pressure metamorphic series.

The metamorphism of the first series may start from very low-grade prehenite-pumpellyite facies, to amphibolite facies and finally reach to the highest granulite facies. Or it may start from very low grade zeolite facies, goes through prehenite-pumpellyite facies, greenschist facies and upto a maximum of amphibolite facies. This is typical of Barrovian metamorphism.

In high pressure series, metamorphism may start from blueschist to greenschist and finally to eclogite facies or from blueshist facies to eclogite facies. Therefore, considering their importance both series are dealt separately here.

10.3.1 Low P-T-Zeolite to Prehenite-Pumpellyite Facies

The systematic study by Coombs et al. in Taringatura hills of Southland in New Zealand and subsequently in the nearby Otama complex, show the presence of systematic very low-grade metamorphic zoning pattern (Fig. 4.2). These areas consist of volcanic tuffs of basic, andesite, dacite and even rhyolite compositions, together with volcanic glass and with high temperature igneous minerals olivine, pyroxene, etc. These rocks are probably metamorphosed to very low-grade condition due to burial. The rocks are not deformed but show series of mineralogical changes while going towards depth. They seemed to have undergone substantial changes in the bulk composition and produced new minerals by the circulating fluids. The volcanic tuffs and glassy matrix are converted to zeolite minerals—heulandite, analcime and stilbite, clay minerals—smectite, kaolinite and montmorillonite, secondary quartz, and carbonate minerals. The coarse grained primary igneous minerals in the tuffs are unaffected during this process. These types of changes are attributed as hydrothermal metamorphism.

The upper part of sequence shows the formation of heulandite, secondary quartz, fine grained phyllosilicates, montmorillonite and celedonite. In the next zone, with increasing depth, calcic plagioclase is replaced by albite, heulandite by laumontite, chlorite and low temperature quartz to high temperature adularia. Still at greater depth, laumontite is replaced by prehenite-pumpellyite with minor epidote. However, it may be mentioned that heulandite, analcime and stilbite are stable phases even during digenesis. Formation of laumontite suggests the beginning of metamorphism or the lower temperature limit of metamorphism. Mineral wairakite (zeolite group) forms slightly at higher temperature than that of laumontite temperature. With further increase in temperature, laumonite is replaced by lawsonite.

Zeolite facies is divided into shallower heulandite-analcite zone and deeper heulandite-albite zone (Fig. 10.1).

Mineral assemblage in this facies include

Chlorite – heulandite/laumontite – calcite;

with these, albite-analcite may be present in Al-rich rocks

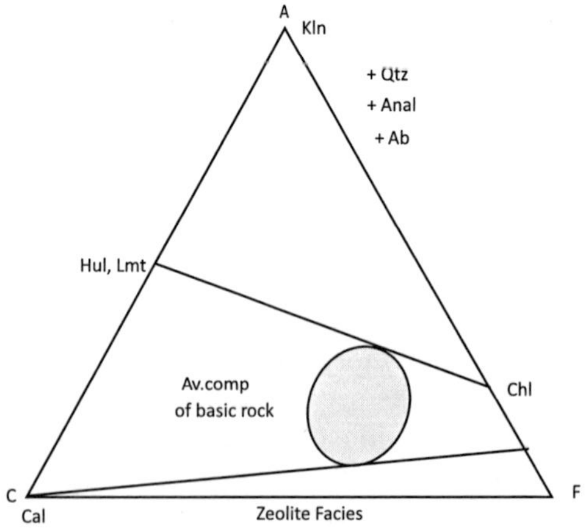

Fig. 10.1: Zeolite Facies Assemblage—Hul/Lmt-Chl-Cal

In the higher temperature side of zeolite facies, lawsonite is replaced by prehenite-pumpellyite minerals that mark the beginning of prehenite-pumpellyite facies at about 200°C and 1 to 3 kb (Fig. 10.2). Other minerals include albite, phengite, lawsonite, chlorite and stilpnomalene and titanite. Higher temperature side of prehenite-pumpellyite facies is marked by the replacement of prehenite to actinolite.

Prehenite-pumpellyite facies mineral assemblages –

Chlorite–prehenite–calcite assemblage at lower temperature

Pumpellyite–actinolite–chlorite–calcite assemblage at high temperature

Pumpellyite disappears at 270-300 °C, and replaced by

Actilonite – epidote – chlorite – calcite assemblage.

The reactions that are responsible for the formation of zeolite minerals, prehenite-pumpellyite and other low temperature minerals are discussed in Winkler (1976) and one may refer for the details. The zeolite and prehenite-pumpellyite facies mineral assemblages are shown in the ACF diagram (Figs. 10.1 and 10.2).

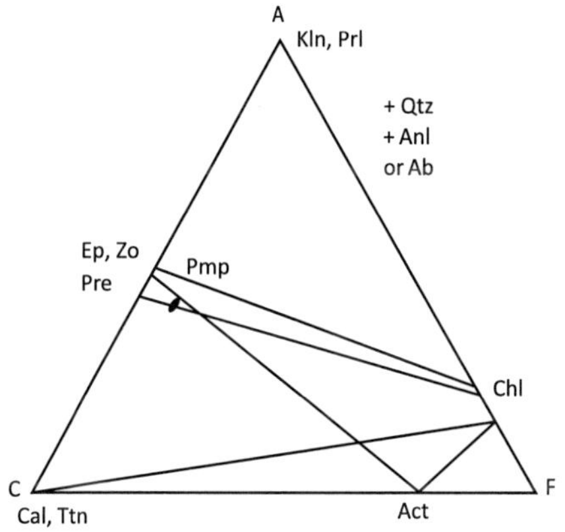

Fig. 10.2: Prehenite-Pumpellyite Facies Assemblages

In many terrains, such type of igneous rocks may be present but not affected by very low-grade metamorphism. It has been suggested that it may be because of low fluid activities in such terrains. Sometimes, such type of volcanic terrains may show the beginning of low grade or greenschist facies metamorphism.

10.3.2 Medium P-T-Greenschist to Amphibolite Facies

The basic rocks in many metamorphic terrains indicate that the rocks have undergone greenschist facies to amphibolite facies and granulite facies metamorphism. The greenschist facies mineral assemblages include chlorite, actinolite, calcite and epidote; or chlorite, epidote, chloritoid assemblage in iron rich rocks (Fig. 10.3).

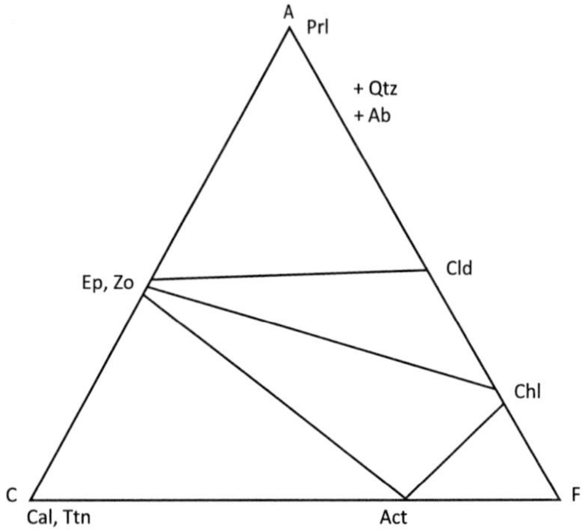

Fig. 10.3: Greenschist Facies Assemblages

The presence of these minerals impart green colour to the rock. Thc rcaction that produces actinolite and epidote is given by

Chlorite + calcite → epidote + actionolite + (CO_2 – H_2O) fluid (10.1)

Equilibrium mineral assemblages as given in Fig. 10.3

Actinolite – chlorite – epidote

Epidote – chloritoid – chlorite

Epidote- actinolite- calcite or titatnite

While going from greenschist facies to amphibolite facies, the following mineralogical changes are observed.

1. The albite (An_{0-10}) is changed to oligoclase (An_{10-30}) composition.
2. Actinolite is changed to hornblende.
3. There is decrease in abundance of actionolite, chlorite, epidote.

4. There is increase in abundance of hornblende, garnet and Ca-plagioclase.
5. Stilpnomelane and albite disappear.

Some of the hornblende forming reactions are:

Chlorite + Al-epidote + quartz → Tschermakite hornblende + anorthite + fluid (10.2)

Albite + actionolite → edenite hornblende + quartz (10.3)

Actionolite + Chlorite + quartz → Al-epidote + Tschermakite hornblende + fluid (10.4)

Hornblende takes more aluminum, sodium and potassium; oligoclase and andesine form at the expanse of epidote; such changes take place at higher P-T condition. However, actinolite to hornblende conversion takes place at lower temperatures than the temperature for albite to oligoclase change.

The amphibolite facies minerals are oligoclase, hornblende and lesser amount of garnet, pyroxene and biotite. Basic rocks metamorphism to amphibolite facies (Fig. 10.4) corresponds to upper garnet zone greenschist facies to staurolite- kyanite/sillimanite zone of amphibolite facies in pelitic rock. Since hornblende can accommodate many elements in variable concentrations, it persists from amphibolite to lower granulite facies condition.

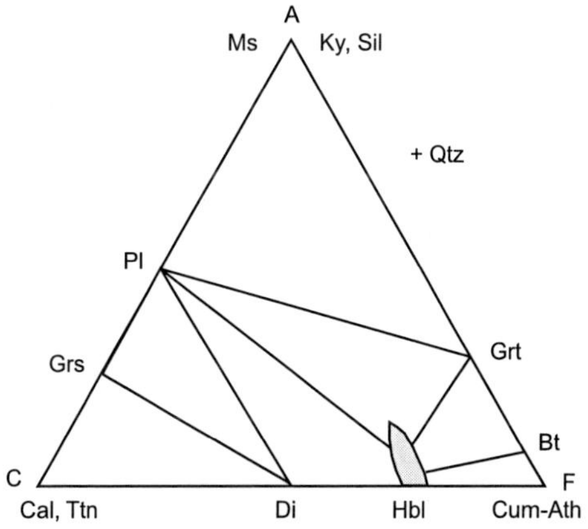

Fig. 10.4: Amphibolite Facies Mineral Assemblages—Hbl-Grt-Plag; Grs-Di-Plag; Di-Plag-Hbl

In the amphibolite to upper amphibolite facies conditions, the basic rocks show formation of certain minerals with variation in whole rock composition and are as follows:

Al-Fe rich and Ca-poor rocks show formation of pyralspite garnet;

Al-poor and Ca-rich rocks show development of clinopyroxene;

Mg-Fe rich and Ca-Al-poor rocks show formation of cummingtonite;

High Mg and low Ca rocks show development of cordierite and anthophyllite

10.3.3 Granuite Facies

Granulite facies rocks are generally dry and show evidence of hot crustal conditions. They are found in the deeply buried continental crust of Pre-Cambrian terrains which occur today as the eroded part of the buried crust. Amphibolite to granulite facies transition takes place at temperatures and pressures of 650 to 800°C and 6 to 9 kb. In general, the rocks show depletion in large ion lithophile (LIL) and incompatible elements as compared to amphibolite facies rocks.

This facies is characterized by anhydrous minerals clinopyroxene, orthopyroxene, plagioclase and quartz. Garnet is common with small amounts of hornblende and/or biotite (Fig. 10.5).

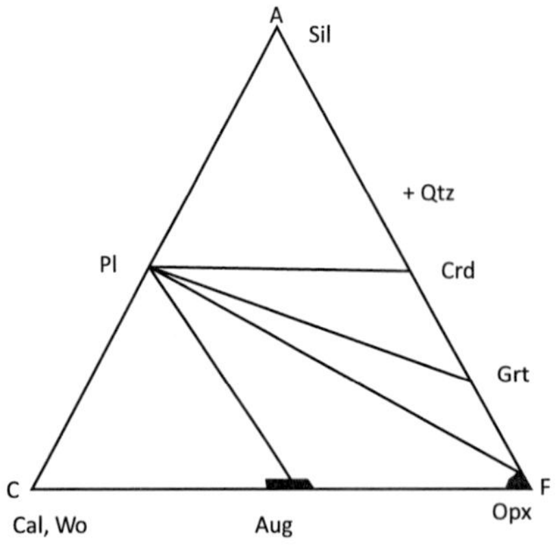

Fig. 10.5: Granulite Facies Assemblages—Plag-Aug-Opx; Opx-Plag-Grt; Grt-Plag-Crd

The presence of hornblende indicates lower temperature side of granulite facies. While going towards higher temperature, hornblende is replaced by pyroxene and plagioclase. Cummingtonite in amphibolite facies break down to orthopyroxene at lower temperature that is lower than hornblende breakdown reaction. Garnet is not present in low pressure granulite facies, but forms at higher pressures by orthopyroxene-plagioclase reaction. The granulite facies mineral assemblages are:

1. Bluish green hornblende + plagioclase + garnet + epidote or clinozoisite + ilmenite
2. Bluish green or greenish brown hornblende + plagioclase + clinopyroxene + garnet + ilmenite
3. Greenish brown hornblende + plagioclase + clinopyroxene + orthopyroxene + ilmenite

Low pressure granulites contain orthopyroxene + clinopyroxene + plagioclase, and the more basic varieties olivine + plagioclase. The formation of hypersthene is explained by several reactions, depending on the minerals in amphibolite facies rock

Hornblende + quartz → hypersthene + clinopyroxene + plagioclase + H_2O (10.5)

Hornblende + biotite + quartz → hypersthene + K-feldspar + plagioclase + H_2O (10.6)

Hornblende + almandine + quartz → hypersthene + plagioclase + H_2O (10.7)

Medium pressure granulites are characterized by garnet + clinopyroxene + orthopyroxene + plagioclase, and also hornblende (Fig. 10.6). Quartz is an accessory, but do not coexist with garnet and clinopyroxene. A well-known example of medium pressure granulites is the Scourian granulites of north-west Scotland.

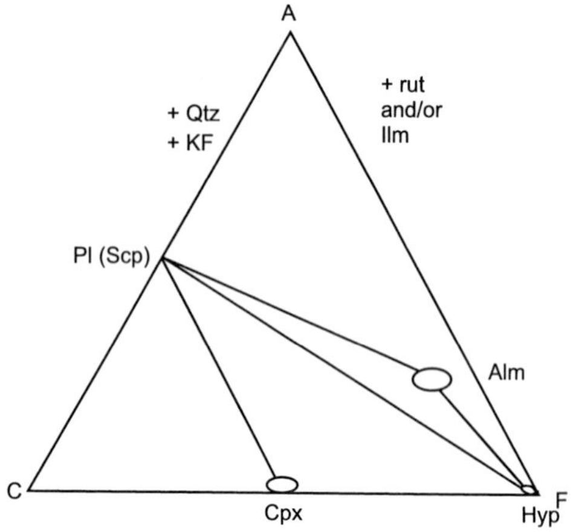

Fig. 10.6: Medium Pressure Granulites. Mineral Assemblages—Pl-Cpx-Hyp; Hyp-Pl-Alm

10.4 High P-T Granulites

The formation of garnet in medium to high pressure granulites is explained by following reactions:

Olivine + plagioclase → garnet (10.8)

Orthopyroxene + plagioclase → garnet + clinopyroxene + quartz (10.9)

Orthopyroxene + plagioclase → garnet + quartz (10.10)

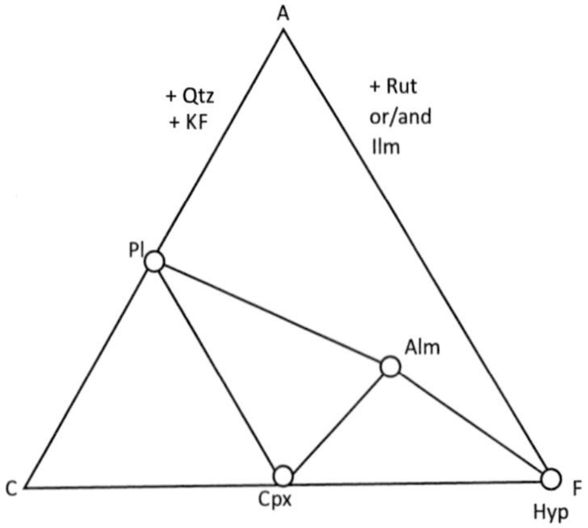

Fig. 10.7: High Pressure Granulites. Mineral Assemblages—Cpx-Alm-Pl; Cpx-Alm-Hyp

These reactions take place when load pressure exceeds a certain value at constant temperature. These reactions are continuous over a P-T range. Experimental investigations reveal that orthopyroxene-plagioclase breaks down to garnet and clinopyroxene at ~700°C and 8 to 10 kb pressure (Fig. 10.7). Beyond this pressure, plagioclase is totally unstable and disappears in eclogite facies.

10.5 High Pressure Facies Series

Basic rocks have a distinct mineralogy at high P-T conditions, as compared to pelitic rocks. The blueschist facies and eclogite facies are formed at higher pressures and relatively at lower to medium temperatures. Blueschist facies rocks are rare in pre-Cambrian subduction zones. This is attributed either due to high thermal gradient in pre-Cambrian period or overprinting of subsequent metamorphism.

10.5.1 Zeolite-Blue Schist Facies

The high pressure series may start with zeolite facies (Fig. 10.1) or blueschist facies (Fig. 10.8). If it starts with zeolite facies and going to blueschist facie, due to increase in pressure, there is a transitional zone, in which the high pressure mineral lawsonite forms between 3.0 to 6.0 kb and < 400°C by the following reactions.

Prehenite → lawsonite + quartz – lower P ~4.0 kb (10.11)

Tremolite + albite + chlorite → lawsonite + glaucophane – higher P > 5.0 kb (10.12)

However, this pressure is not enough to produce glaucophane or jadeite. In case the high pressure series starts with blueshist facies (Fig. 10.8), it might have followed any of the following paths depending on the P-T gradients.

Zeolite facies – prehnite – pumpulite facies – blueschist facies or

Zeolite facies – prehnite – pumpellyite facies – greenschist facies – blueschist facies

The rock is blue colored due to the presence of blue colored Na-amphibole glaucophane and termed as blueschist. Several reactions produce glaucophane and some of the reactions are:

Pumpellyite + chlorite + albite → glaucophane + epidote + water (10.13)

Tremolite + chlorite + albite → glaucophane + epidote + water (10.14)

Tremolite + chlorite + albite → glaucophane + lawsonite (10.15)

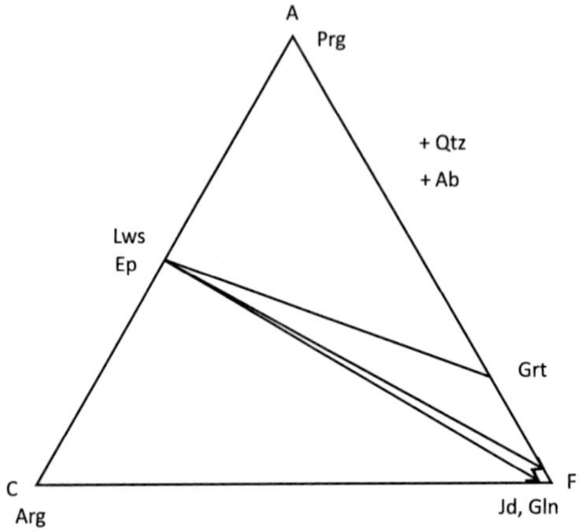

Fig. 10.8: Blueschist Facies Assemblage—Lws/Ep-Jd/Gln-Grt

All these reactions are sensitive to pressure. At still higher pressures, albite breaks down to jadeite and quartz.

Albite → jadeite + quartz (10.16)

If jadeite is present without quartz, it indicates relatively low pressure blueschist facies condition. If jadetite and quartz are present, they indicate relatively higher pressure blueschist facies.

The assemblage glaucophane-epidote indicates low pressure, and glaucophane-jadeite-quartz indicates high pressure. Chance presence of calcite at low pressure may be converted to aragonite at high pressure.

The glaucophane and paragonite assemblage of bluescist facies reacts and produces omphacitic pyroxene at > 12.00 kb. If there is no paragonite in blueschist facies, glaucophane remains stable in eclogite facies.

Transitional blueschists to greenschist facies reactions are:

Blueschist facies to greenschist facies

Glaucophane + lawsonite → albite + chlorite + actionolite (10.17)

Glaucophane + zoisite + quartz→ albite + chlorite + actinolite + H_2O (10.18)

Crossite + epidote → albite + chlorite + actinolite + Fe-oxide + H_2O (10.19)

Transitional blueschist to eclogite facies reactions are:

Blueschist facies to eclogite facies

Zoisite + glaucophane → garnet + omphacite + paragonite + quartz + H_2O (10.20)

Lawsonite + albite → zoisiite + paragonite + quartz + H_2O (10.21)

Lawsonite + jadeite → zoisite + paragonite + H_2O (10.22)

10.5.2 Eclogite Facies

Eclogite occur as xenoliths in kimberlites or basalts (e.g. Oahu, Hawaii), as bands or lenses in migmatitic gneisses (e.g. west Norway) and as bands or lenses associated with blueschist (e.g. New Caledonia, Franciscan terrane of California, Alpine – Tethyan chain).

The eclogites are classified into low, medium and high temperature eclogites. Eclogite parent rock is mainly pillow lavas, but not mantle derived magmatic rock.

Eclogite facies is marked by the appearance of omphacite pyroxene and pyrope-almandine garnet (Fig. 10.9).

At eclogite formation pressure, plagioclase is not stable and breaks down to produce either glaucophane or jadeite. Blueschist to eclogite facies transformation is by the reaction

glaucophane + paragonite → pyrope garnet + jadeite + quartz + H_2O (10.23)

This reaction takes place at a pressure of >12 kb. At still higher geothermal gradients, amphibolite or granulite facies mineral assemblages may produce eclogite facies assemblage. Plagioclase reacts with amphibole, pyroxene or olivine and produce omphacite and pyrope-grossular garnet. An-rich plagioclase breaks down mark the beginning of eclogite facies metamorphism (Fig. 10.9). Plagioclase may be converted to kyanite, quartz and garnet by the reactions:

An-plagioclase + H_2O → Kyanite + zoisite + quartz (10.24)

An-plagioclase → Grossularite garnet+ kyanite + quartz (10.25)

An-plagioclase → Ca-Tschermakite + quartz (10.26)

If paragonite is present in the blueschits facies rock, it breaks down to jadeite and kyanite by the reaction

Paragonite → jadeite in omphacite + kyanite + fluid (10.27)

Low temperature eclogites contain zoisite, epidote, amphibole, phengite, paragonite and pyrope garnet. Medium grade eclogites occur in high-grade gneissic terrain and contain quartz, zoisite, paragonite, calcite-amphibole and pyrope-rich garnet. These are developed due to thrust stacking where the pressure exceeds ~20.0 kb. High temperature eclogites are derived from subducted oceanic crust or de-laminated crustal under-plates incorporated in the mantle. Important minerals include the presence of kyanite, coesite and micro diamond.

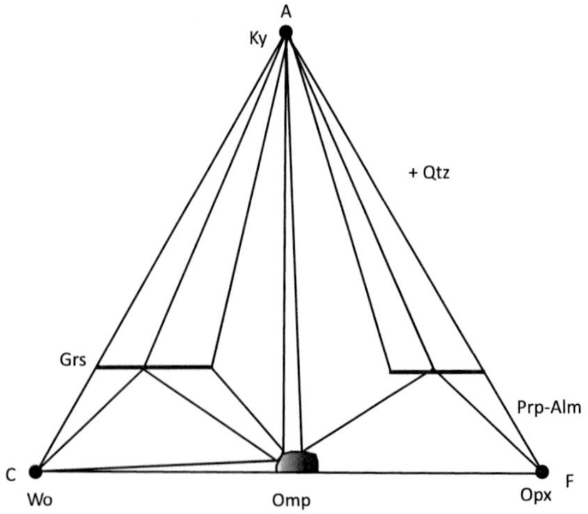

Fig. 10.9: Eclogite Facies Assemblage

Basic rock – omphacite – pyrope almandine garnet;

Omphacite – grossularite – wollastonite

Kyanite – coesite (quartz) – omphacite

Pyp – alm garnet – kyanite – quartz

10.6 Ultra High Pressure-Temperature Metamorphism

To understand the behavior of continental crust during subduction, continental collision and crustal thickening, it is necessary to study UHP and UHT metamorphism of the rocks that occur in the tectonic zones.

10.6.1 Ultra High Pressure (UHP) Metamorphism

UHP metamorphism is developed in mafic and ultra-mafic rocks at temperatures and pressures of 700 to 950°C and 27-50 kb (2.7 to 5.0 GPa). Presence of mineral coesite inclusions in the garnets of white micaschist is indicative of UHP metamorphism. About 20 such UHP metamorphic localities are reported in the world. Generally, the UHP rocks occur as long and linear intermittent belts, produced by continental subduction and/or collision. These UHP rocks are developed at a depth of about 150 to 200 kms. The UHP rocks are inter-layered with quartzite, marble, micaschist, quartzo-feldspathic gneiss, limestone, shale and greywacke. These rocks are characteristic of continental fore arc wedge.

10.6.2 Ultra High Temperature (UHT) Metamorphism

UHT metamorphism is equivalent to medium to high pressure granulite facies metamorphism. Peak temperature is about 900 to 1100°C and pressure is about 7 to 13 kb (0.7 to 1.3 GPa). About 40 localities occur all over the world. UHT metamorphism is common in pre-Cambrian terrains, where heat flow was higher. Crustal thickening and rising of the mantle possibly supplied additional heat to produce such UHT metamorphism. The mineral assemblages are sapphirine-quartz; orthopyroxene – sillimanite- osumullite; wollastonite – scapolie – quartz. Orthopyroxene contains more than 2 to 8% of Al_2O_3. Coexistence of garnet – sillimanite – sapphirine indicates UHT metamorphism.

10.7 Calc-Silicate Rocks

Calc-silicate rocks are similar in composition to basic-ultramafic rocks. Such calcareous rock with higher clastic sediments produce Ca-Mg-Fe-Al-K-Na silicate minerals diopside, epidote group, grossular garnet, Ca-amphibole, plagioclase, vesuvianite, magnetite, scapolite, together with muscovite, biotite and chlorite in medium to high grade metamorphism. However, in water-rich fluid composition calc-silicates vesuvianite, margarite and grossular are more stable than other calc-silicate minerals that are common in CO_2-rich fluids condition. Presence of additional Al_2O_3, FeO, Na_2O, K_2O in the rock produce more complex mineralogy due to metamorphism. There is no systematic mineralogy for calc-silicate rocks, as observed in basic or pelitic rocks, due to variable whole rock chemical and mineralogical composition from one terrain to another terrain, and therefore, difficult to make any comparison.

So based on field study in two terrains of USA, researchers have described five zones for calc-silicate rocks. They are (a) ankerite zone, (b) biotite zone, (c) amphibole zone, (d) zoisite zone and (e) diopside zone from low to high grade. These zones may or may not be present in other calc-silicate rocks of other metamorphic terrains.

10.7.1 Ankerite Zone

Lowest grade – Mineral assemblages include ankerite, quartz, albite, muscovite, ± chlorite and ±calcite. Dolomite with ~20% (Fe +Mn) is known as ankerite.

10.7.2 Biotite Zone

It is formed by the reaction

Muscovite + quartz + ankerite → calcite + chlorite + biotite + CO_2 (10.28)

This reaction takes place at the same grade as in normal pelites. At lower grade of this zone biotite is ~3% and increases to ~20% on the higher temperature side of this zone. Muscovite decreases and ankerite is out at higher grade. Plagioclase is An_{29} in this zone.

10.7.3 Amphibole Zone

Actinolite-tremolite (Ca-amphibole) is formed by reaction

Calcite + chlorite + quartz + albite → Actinolite + anorthite + H_2O + CO_2 (10.29)

Mineral assemblage in this zone is calcite, amphibole, quartz, plagioclase, biotite ± chlorite. Tremolite-actinolite occur in lower temperature side and becomes more Al-rich across this zone. Average plagioclase composition is An_{70}.

10.7.4 Zoisite Zone

Zoisite forms as reaction rim around plagioclase and at the contact of calcite. Mineral assemblage include zoisite, Ca-amphibole, quartz, plagioclase (An_{74}), calcite, ± biotite and ± microcline. Microcline is formed by reaction

Biotite + calcite + quartz → Ca-amphibole + microcline (K-fedspar) + H_2O + CO_2 (10.30)

Most prograde reactions produce H_2O + CO_2 due to which CO_2 buffereing is very high in the fluid. In case H_2O buffereing is high in the fluids, vesuvianite and grossularite are formed. High H_2O content is possibly contributed by granitic intrusions in USA area. Variation in H_2O + CO_2 ratio of the fluid is attributed to associated pelitic-calcareous sediments and granitic plutons.

10.7.5 Diopside Zone

This zone is equivalent to sillimanite zone in pelitic rocks. Diopside is formed by continuous reaction

Tremolite + calcite + quartz → Diopside + H_2O + CO_2 (10.31)

Mineral assemblage in diopside zone includes zoisite, Ca-amphibole, calcite, quartz, plagioclase (An_{79}), ± biotite and ± microcline. Therefore, based on whole rock composition, there may be more number of metamorphic zones in calc-silicate rocks, besides the zone described above.

10.8 Ultramafic Rocks

Dunite and Peridotite group rocks are called as ultranafic rocks and they contain olivine, orthopyroxene, clinopyroxene and sometimes with garnet. They are called as Alpine or orogenic peridotites. They also occur in ophiolite suite of rocks in young subduction zones. Peridotite is retrogressed to serpentinite before or during collisional event. Later such rocks undergo metamorphism producing high grade mineral assemblages and be plotted in CaO-MgO-SiO_2 (CMS)-H_2O system. FeO and Al_2O_3 are in less wt% and have minor effect during metamorphism.

Low metamorphic grade is dominated by serpentine group of minerals lizardite, crysotile and antigorite. Below 500°C, olivine and pyroxene is converted to serpentine, and Fe released from such minerals is changed to magnetite. Formation of serpentine minerals, i.e. asbestos minerals, starts with the sequence lizardite, lizardite + crysotile, crysotile + antigorite to antigorite from low to high grade metamorphism. Olivine is replaced by mesh type lizardite and crysotile, orthopyroxene is psudomorphosed to bastite. Antigorite at higher grade is coarse grained. Crysotile breaks down at 250°C to antigorite by reaction

Crysotile → antigorite + brucite (10.32)

A traverse along medium P-T path in one such peridotite terrain has shown the following mineralogy in different metamorphic grades. Peridotite has olivine: ortho-pyroxne: clinopyroxene in 6:3:1 ratio. In greenschist facies condition, peridotite has pure serpentine with some brucite and diopside. Talc and tremolite are present in SiO_2 rich serpentinites. Diopside is stable in lower grades when the fluid is pure H_2O, whereas antigorite variety of serpentine breaks through several reactions and one such reaction is

Antogorite + brucite → Fosterite + H_2O (10.33)

This reaction marks the high temperature stability of brucite. Olivine formed in prograde reaction is Mg-rich than the olivine of the normal peridotite. Another reaction in which antigorite and diopside reacts at ~530°C to

Diopside + antigorite → fosterite + tremolite + H_2O (10.34)

Tremolite is found in low SiO_2 ultramafic rocks and more SiO_2 rich assemblage contains both tremolite and diopside. Antigorite is finally goes out by reaction

Antigorite → Fosterite + talc + H_2O (10.35)

Appearance of talc is characteristic of olivine bearing ultramafics. Olivine formed in talc-rich matrix shows spinifix texture.

At pressure above 1.5 GPa, antigorite breaks down to enstatite and/or fosterite by reactions

Talc + antigorite → Enstatite + H_2O (10.36)

Antigorite → Fosterite + enstatite + H_2O (10.37)

At ~650°C, talc is out and enstatite is formed by reaction

Talc + fosterite → enstatite + H_2O (10.38)

This reaction is the low temperature stability limit of enstatite formation in amphibolite facies. Enstatite bearing ultramafic rocks is associated with amphibolite in the field. At low pressures, anothophyllite is introduced by reaction

Talc + fosterite → anthophyllite + H_2O (10.39)

Talc + enstatite → anthophyllite (10.40)

Anthphyllite has limited stability, in which upper temperature limit and under 0.5 GPa pressure, it produces enstatite by reaction

Anthophyllite + forsterite → enstatite + H_2O (10.41)

Anthophyllite stability is larger in SiO_2-rich ultramafics, but such type of ultramafic is rarely found. Talc is more abundant than fosterite in SiO_2-rich ultramafic rocks and it is stable upto 750°C. Finally, talc disappears by reaction

Talc → anthophyllite + quartz + H_2O (10.42)

At higher pressure talc gives rise to

Talc → quartz + enstatite + H_2O (10.43)

At ~800°C, diopside reappear in typical ultramafics by reaction

Fosterite + tremolite → enstatite + diopside + H_2O (10.44)

The mineral assemblages are equivalent to granulite facies metamorphism. Anthophyllite breaks down at ~800°C by reaction

Anthophyllite → quartz + enstatite + H_2O (10.45)

At 900°C, in medium P-T path, tremolite in SiO_2-rich rock breaks down to

Tremolite → enstatite + quartz + H_2O (10.46)

The conversion of peridotite to serpentine group minerals, diopside, talc, tremolite, etc. depends entirely on the ratio of minerals in ultramafic rocks, fluid activity in the respective terrain and P-T condition of metamorphism. Therefore, one has to take utmost care in studying the ultramafic metamorphism.

Chapter 11

Metamorphism of Pelitic Rocks

The fine grained clastic sediments, derived due to weathering and erosion of the continental crust, accumulated at the slopes of passive and active continental margins is known as pelitic sediments. They grade in to shales, mudstones, sandy sediments and greywackes, while going from continental slope to the source region. Pelites are dominated by Al-K rich clays montmorillonite, kaolinite, smectite, etc., fine grained white mica sericite, phengite, chlorite, quartz, albite, K-feldspar, iron oxides, varieties of hydra-oxides, zeolites, carbonates and organic matter. Clay, mica, chlorite and quartz occur predominantly with minor amounts of other minerals. The sediments usually have high Al_2O_3, K_2O, SiO_2 and low CaO contents. Appreciable amount of mica in pelites, led to the formation of slate, phyllite, micaschist and gneiss in prograde metamorphism. The sediments contain nearly ~30% of water and CO_2 that are expelled during diagenesis and metamorphism. Pelites are sensitive to temperature and pressure and show significant mineralogical changes during prograde metamorphism. Based on the classic study of pelites in Scottish Island, the Barrovian metamorphic zones were suggested in the early 20th century (Fig. 4.1). The chemical composition of the pelites can be represented in the six-component system K_2O-FeO-MgO-Al_2O_3-SiO_2-H_2O (KFMASH). As SiO_2 and H_2O are present in most of the minerals, they are eliminated and the remaining three or four components are represented in AKF or AKFM (AFM) diagrams. Since muscovite is present upto amphibolite facies and K-feldspar in upper amphibolite to granulite facie rocks, muscovite or K-feldspar is used as projection point to plot all the minerals in AFM diagram. In Fig. 11.1, muscovite is used as projection point and the peltic minerals are shown (Fig. 11.1). The reactions written are not balanced reactions and are only to show how the new minerals are formed from the reactions.

11.1 Diagenesis and Low Grade Metamorphism

Shale becomes an argillite (massive rock) or slate (foliated rock) during diagenesis and low grade metamorphism. The boundary between diagenesis and very low grade metamorphism is gradational. It is difficult to delineate the onset of metamorphism in this rock. During diagensis, clay minerals are changed to pyrophyllite, chlorite, and thermally stable illite clay. Illite shows poor crystallinity in diagenetic stage, and with increasing temperature it becomes an ordered crystalline structure. Based on illite crystallinity, it is possible to define diagenesis and low-grade metamorphism. Carbonaceous matter present in pelites is changed to graphite during diagenesis and low-grade metamorphism. Graphite vitrinite reflectance (when polarized light falls on the polished mineral graphite, as it is not transparent like silicates but opaque, certain amount of light is reflected back which is called as reflectance) increases with increasing metamorphism. Like illite crystallinity, vitrinite reflectance of graphite can also be used to decipher the grade of metamorphism. Slate contains illite, sericite, phengite, quartz, chlorite, albite, K-feldspar, minor carbonates and hematite. Pyrophyllite is present in slates and phyllites, only when K-feldspar and albite are absent. Pyrophyllite and phengite is present in some slates. Pyrophyllite is produced by the reaction.

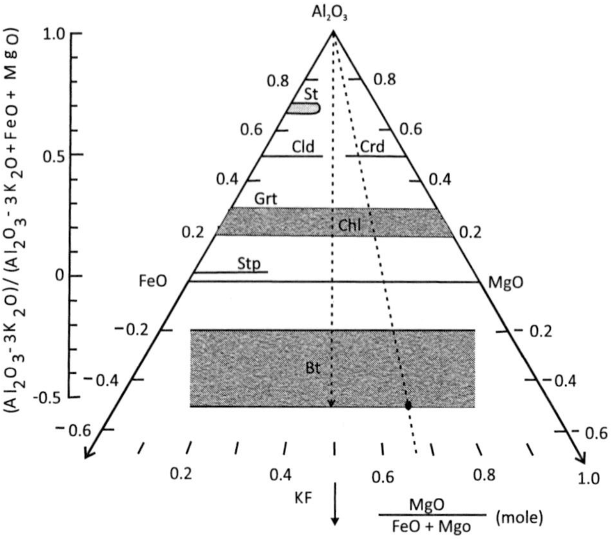

Fig. 11.1: AFM Projection from Muscovite for Pelitic Minerals (after J.B Thompson)

Kaolinite + 2 quartz → pyrophyllite + water (11.1)

This reaction takes place at 325°C/1 kb to 375°C/4 kb. This does not indicate very low-grade metamorphism. However, this reaction may take place at very low grade condition, provided graphite is present in carbonaceous clay or black shale rock. This is due to the release of CO_2 from carbonaceous matter, H_2O pressure is reduced, and hence the reaction takes place at very low temperature.

If the rock has Na-montmorillonite or pyrophyllite or kaolinite with albite, paragonite is formed by reactions

Na-montmorillonite + albite → paragonite + quartz (11.2)

Kaolinite/pyrophyllite + albite → paragonite + quartz + water (11.3)

Paragonite occurs together with phengite, chlorite, quartz, sometimes with calcite/dolomite. Paragonite once formed in very low grade continues upto medium grade or amphibolite facies condition. If paragonite and carbonates are present in low grade, it reacts and produces plagioclase, margarite, zoisite/clizozoisite in the rock. No other mineralogical changes are observed in very low-grade metamorphism. Sericite is hydrous pyrophyllite with high SiO_2, Na_2O, MgO, H_2O composition. Phengite and celedonite are solid solutions of muscovite. Na-rich muscovite is called paragonite found in high Al-rich rocks, Ca-rich muscovite called as margarite is found in marly rocks (carbontes + pelites), and true muscovite in K-rich normal pelites.

11.2 Greenschist Facies

11.2.1 Chlorite Zone

Barrovian zone metamorphism starts with chlorite zone in the temperature range of 350 to 400°C. Pelite composition can vary from low-Al to high-Al varieties and contain phengite, muscovite, chlorite, quartz and albite minerals, and in semipelites there may be additional k-feldspar and minor calcite. As mentioned earlier, the grade can be determined by the use of illite crystallinity and graphite reflectance. The minerals of this zone are given in AKF and AFM diagrams (Fig. 11.2).

11.2.2 Biotite Zone

This zone is characterized by the formation of biotite. The mineral assemblages include biotite, chlorite, muscovite, albite and quartz. Biotite is formed by a number of reactions depending on the mineralogy of the pelites and greywackes. Greywackes may contain K-feldspar and chlorite in the lower grade. These two minerals react and produce biotite by the reaction

K-feldspar + chlorite → biotite + muscovite + quartz + water (11.4)

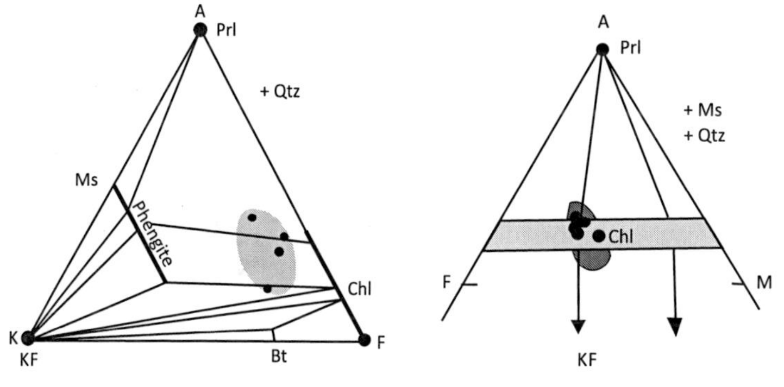

Fig. 11.2: AKF and AFM Diagrams for Lower Greenschist Facies - Chlorite Zone. Black dots indicate the av.composition of pelites. Assemblages: Prl-Phen-Kf, KF-Chl-Phen

This discontinuous reaction takes place at the higher temperature side of the chlorite zone and not on the actual biotite grade temperature of the Barrovian zone. However, this reaction goes to completion in a narrow temperature range. Normal pelites do not contain K-feldspar; instead contain plenty of phengite-muscovite and chlorite. These two minerals produce biotite by the continuous reaction

Phengite + chlorite → biotite + phengite poor muscovite + quartz + water (11.5)

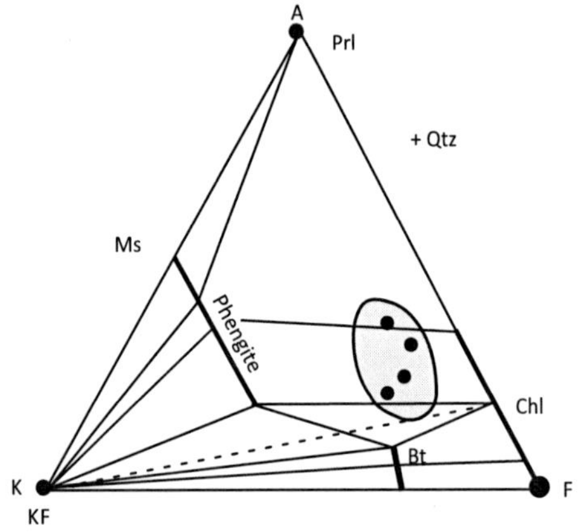

Fig. 11.3: AKF Diagram for Pelites-Biotite Zone. Assemblages KF-Bt-Phen Ms; Ms-Chl-Bt

With high phengite white mica and low Mg-chlorite with low Mg contents, this reaction starts at the beginning of biotite grade (Fig. 11.3) and continues to garnet zone producing more pure muscovite at higher temperatures. Chloritoid is another mineral found in this zone, but in Fe-rich pelites. It is produced by the reaction

Chlorite + pyrophyllite → chloritoid + quartz + water (11.6)

However, chloritoid occur at a temperature below 250°C in Fe system (Mg-free). This is shown in Fig 11.4b, where chloritoid plots between pyrophyllite and chlorite on the A-F side. It is suggested that appearance of chloritoid may be on upper chlorite or lower biotite zone temperature. Another chloritoid forming reaction in Fe-rich rocks

Fe-chlorite + muscovite → Fe-chloritoid + annite + quartz + water (11.7)

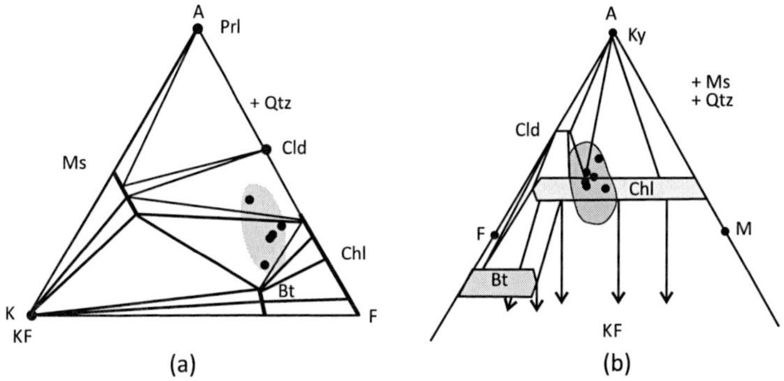

Fig. 11.4: AKF (a) and AFM (b) Diagrams for Biotite Zone. The Assemblages—KF-Ms-Bio, Bio-Ms-Chl, Ms-Chl-Ctd

If chloritoid is present, almandine garnet may be formed by the reaction

Fe-chloritoid + annite + quartz → almandine + muscovite + water (11.8)

This reaction is possible only in Mg-free system, but such system is not found in natural pelites. Garnet may also be produced by another chloritoid reaction

Chloritoid + quartz + water → garnet + chlorite + muscovite (11.9)

The appearance of garnet through chloritoid reaction is considered to be garnet isograd or beginning of garnet zone.

11.2.3 Garnet Zone

In this zone, garnet appears along with biotite, muscovite, chlorite, albite, epidote and minor amounts of ilmenite and magnetite. Garnet occurs as porphyroblasts in the mica schist and easily identifiable in the field. Garnet is generally almandine rich in the Barrovian zone metamorphism. Sometimes, spessartine-rich garnet porphyroblasts, smaller in size, occur in Mn-rich pelites, that form slightly at lower temperatures (upper biotite grade) than garnet zone temperature. Garnet is formed by the reaction

Chlorite + muscovite → garnet + biotite + quartz + water (11.10)

As garnet, chlorite, muscovite are solid solutions with variable Fe: Mg ratio, with increase in temperature, the distribution of Fe and Mg between these minerals is as follows: $X_{Mg}^{Chl} > X_{Mg}^{Bio} > X_{Mg}^{Gt}$. Chlorite becomes more Mg-rich from chlorite zone to upper garnet zone. High Mg-rich chlorite continues to be present in staurolite zone. The biotite and garnet zone minerals assemblages are shown in Figs 11.4 and 11.5. It may be noticed that chlorite-biotite tie lines are fixed for a particular bulk composition of pelites and at a specified P-T range. Garnet is formed by continuous reaction initially with Fe-rich composition and progressed to Mg-rich composition at higher temperatures. The same thing is true with biotite and chlorite that becomes more Mg-rich towards higher temperature.

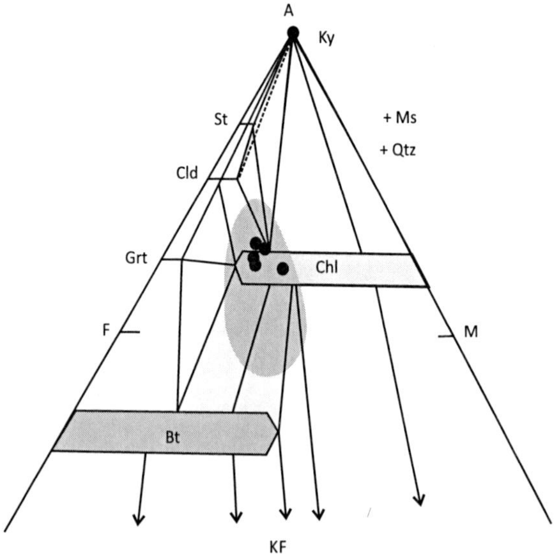

Fig. 11.5: AFM Diagram for Upper Garnet Kyanite Zone Showing Average Pelitic Compositions (black dots). The Assemblages—Grt-Chl-Bt, Grt-Bt-Chl, Ctd-Chl-St (Fe-rich pelite)

In some high Al-rich pelites, chloritoid is found in garnet zone and rarely with biotite. At lower pressures, chloritoid occurs with garnet, muscovite, chlorite, plagioclase, paragonite and quartz. In Fig. 11.5, it may be noticed that biotites do not occur together with chloritoid. At higher temperature side of garnet zone, i.e. transition to staurolite-kyanite grade, plagioclase changes its composition from albite to oligoclase/andesine. In Fe-rich pelites, staurolite may form at the higher temperature side of garnet zone, as shown in Fig. 11.5.

11.3 Amphibolite Facies

11.3.1 Staurolite-Kyanite Zone

Staurolite forms in Fe-rich, Al-rich and Ca-poor pelites. Since the available Al is used to form staurolite, Al_2SiO_5 is negligible or scarce in such rocks. If chloritoid is present in the assemblage, staurolite is formed by the reaction

Chloritoid + quartz → staurolite + garnet + H_2O (11.11)

Chloritoid + kyanite → staurolite + quartz + H_2O (11.12)

Reaction 11.12 takes place at ~520 to 575°C and 2.0 to 8.0 kb pressure.

Staurolite is formed from the garnet zone mineral assemblages by the following reactions

Garnet + muscovite → staurolite + biotite + quartz (11.13)

Chlorite + muscovite → staurolite + biotite + quartz + H_2O (11.14)

These two reactions (11.13 and 11.14) possibly take place at 540–565°C and 4.0–7.0 kb.

These are discontinuous reactions at a fixed temperature and wide range of pressures. Plagioclase becomes more An-rich by the use of Ca-epidote present in lower metamorphic zones. Chlorite is absent in staurolite zone, except for Mg-rich chlorite. The change in tie lines garnet-chlorite to staurolite-biotite can be seen in Figs. 11.5 and 11.6.

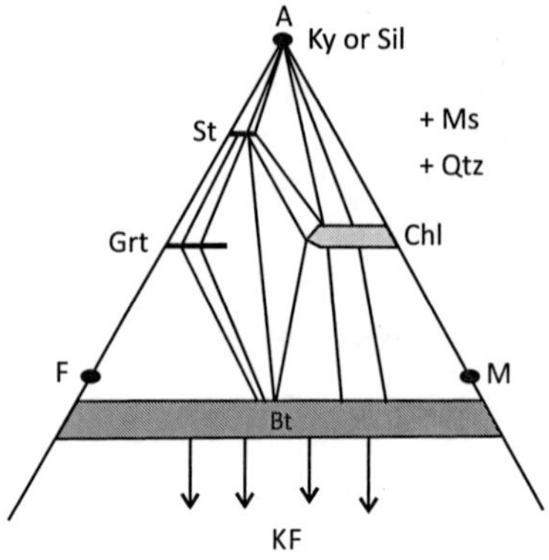

Fig. 11.6: AFM Diagram for Staurolite-Kyanite Zone—Black dots—av.pelitic Compositions. The Assemblages: Grt-St-Bt, St-Ky-Grt, St-Ky-Bt

Kyanite is formed in the same P-T condition as that of staurolite, by the following reactions in relatively Mg-rich rocks.

Muscovite + staurolite + chlorite → kyanite + biotite + quartz + H_2O (11.15)

Staurolite + muscovite+quartz → kyanite+biotite+H_2O (11.16)

11.3.2 Lower Sillimanite Zone or Beginning of Upper Amphibolite Facies

With further increase in temperature from staurolite zone to the beginning of sillimanite zone, staurolite disappears and fibrolite (fine needles of sillimanite) is formed (Fig. 11.7). Kyanite persists to higher grades. Fibrolite is mostly developed within kyanite, garnet, micas, feldspars, etc. Prismatic sillimanite is not found in this zone. Disappearance of staurolite can be written by the reaction

Staurolite + quartz → Garnet + Al_2SiO_5 + H_2O (11.17)

Staurolite + muscovite + quartz → Sillimanite + garnet + biotite + water (11.18)

Reaction (11.17) takes place ~700°C and 4.0 to 7.0 kb.

Muscovite remains stable in this zone, and some muscovite may consume in certain reactions.

In this zone, sillimanite/kyanite is developed by the reaction

Chlorite+muscovite+quartz → kyanite+biotite+H_2O (11.19)

This reaction takes place only when the composition of chlorite is Mg-rich, and chlorite finally disappears with increasing temperature. However, in normal pelites with increasing temperature staurolite is present along with fibrolite/kyanite. Staurolite disappears at higher temperature by the reaction:

Staurolite + chlorite + (muscovite + quartz) → kyanite + biotite + H_2O (11.20)

In this zone, kyanite is converted to sillimanite by polymorphic transformation.

11.4 Upper Sillimanite Zone or Almandine Amphibolite Facies

In this zone, primary muscovite is absent with the presence of plenty of K-feldspar, sillimanite and biotite. Muscovite breaks down to K-feldspar and sillimanite by reaction

Muscovite + quartz → K-feldspar + sillimanite + H_2O (11.21)

This reaction takes place at 650 to 675°C and 3.0 to 4.0 kb. This reaction is important due to the formation of visible sillimanite. This zone is termed as muscovite-out or sillimanite-K-feldspar-in isograd. Muscovite reaction produces more k-feldspar and the rock becomes granular and less in schistosity. The change in mineral assemblage is shown in Fig. 11.8.

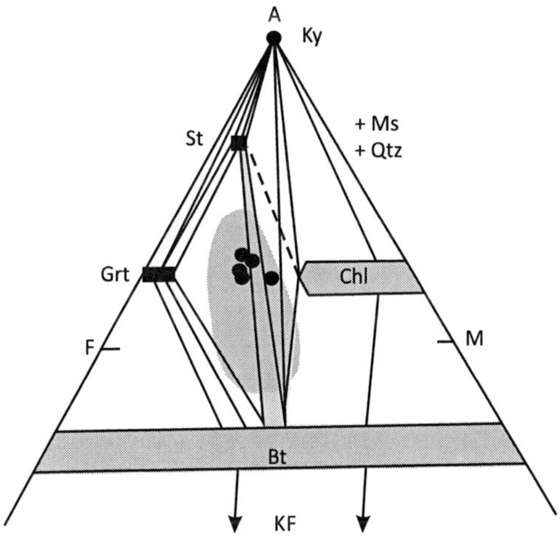

Fig. 11.7: AFM Diagram for Kf-sillimanite Zone for Average Pelitic Compositions and the Assemblages: Sill-Bt-Kf-Qtz, Grt-Bt-Sill-Qtz

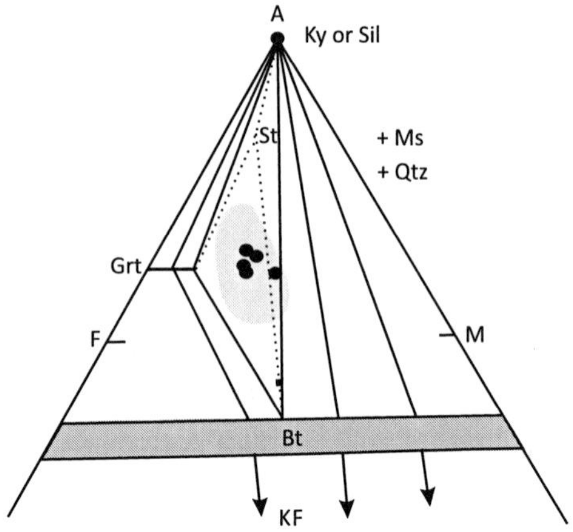

Fig. 11.8: AFM Diagram for Upper Sillimanite or Almandine Amphibolite Facies. Assemblage: Grt-Bt-Sil, Grt-Kf-Bt

At still higher temperature and low pressure cordierite appears in Mg-rich pelites by the reaction

Phlogopite (Mg-biotite) + sillimanite + quartz → Mg-cordierite + K-feldspar + H_2O (11.22)

Cordierite is also produced by another reaction at low pressure

Biotite + sillimanite → garnet + cordierite + H_2O (11.23)

However, in normal pelites garnet-cordierite assemblage is not found in this grade.

11.5 Partial Melting of Pelites

In this isograd, water released by reaction (11.21) may induce partial melting, as the temperature is above wet granite solidus, and migmatisation. If water released in this zone escapes out of the system and rock becomes dry, with further rise in temperature, the rock reaches to granulite facies P-T condition.

Partial melting of pelite is explained with the help of petrogenetic grid, developed by Spear for the pelites (see Winter book for details).

To draw this grid, three assumptions are made, viz.

(a) For melting, water is available only by dehydration reactions and not derived from external source.

(b) In sub-solidus state, dehydration reactions continuously generate vapour/fluid and mostly it leaves the rock. Hence, initial melting may occur in water saturated condition, but to a very limited extent.

(c) Once melting started, fluid released by dehydration reactions dissolve into the melt as long as the melt is not removed.

These assumptions restrict only to a few reactions for vapor saturated melting at lower temperature. Fluids enter into the melt in the beginning stage reduces subsequent reactions to vapor absent condition. The important melting reactions in granulite facies condition are shown in the modified KFMASH (K, Fe, Mg, Al, Si, H_2O) P-T grid (Fig. 11.9). In this grid, the effect of other components such as CaO, Na_2O, MnO are not considered as these are present significantly in less concentration in the pelites.

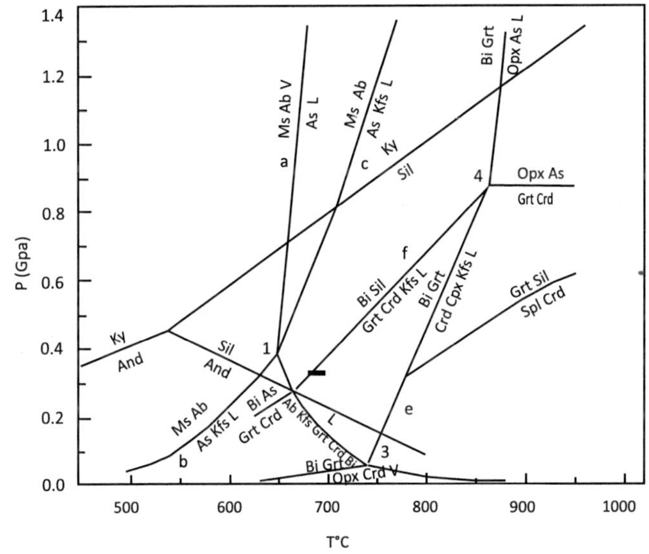

Fig. 11.9: Simplified Petrogenetic Grid for Partial Melting (Migmatization) of Pelitic Rock

In Fig. 11.9, at the medium P-T condition and above invariant point 1, melting occurs due to muscovite dehydration reaction with excess vapour (univariant line a).

Muscovite + albite + quartz + $H_2O \rightarrow Al_2SiO_5$ + liquid (11.24.1)

If albite is absent, this reaction may produce little melt. Below invariant point 1, at low P-T gradient, univariant line (b), muscovite dehydrates below solidus producing k-feldspar and Al_2SiO_5 with no melting. In medium P-T path i.e. ~650 to ~750°C, 4 to 8 kb, little or no melt is generated. Whatever vapour generated in the initial stages is dissolved into the melt, and no free vapor is available in the later stages for further melting, even though muscovite, biotite, albite, Al_2SiO_5, quartz etc., are present (univariant line (c)

Muscovite + albite + quartz $\rightarrow Al_2SiO_5$ + kfeldspar + liquid (b in Fig. 11.9) (11.24.2)

This reaction with 7 to 10% modal muscovite will generate nearly 7 to 20% melt. More melt may be generated by biotite reaction with the formation of garnet, cordierite and melt. This may be end stage of melting, as the product minerals indicate granulite grade condition.

Biotite + sillimanite $\rightarrow$ garnet + cordirite k-feldspar + liquid (f in Fig. 11.9) (11.24.3)

At low P-T (< 650°C and < 4.0 kb), below invariant point 1 no melt is formed. At low pressure (< 4.0 kb) and higher temperature of 650 to 750°C (univariant line d), major vapour saturated melting takes place by reaction,

Kfeldspar + albite + garnet + cordierite + biotite + quartz + $H_2O \rightarrow$ liquid (11.24.4)

Therefore, at low pressure and temperature > 700°C, significant amount of melt is generated.

11.6 Migmatite

The unaltered or slightly altered parent rock is called as paleosome or melanosome. The term neosome is used to for the newly formed migmatite. Neosome comprises two parts, viz. the leucosome and the melanosome. The term migmatite is coined by Sederholm (1907). It is a mixed rock having igneous looking component and metamorphic component. Migmatite is formed in pelitic, sandy arkosic sediments and to some extent in basic rocks with suitable mineralogical composition. Melanosome contains dark-colored minerals biotite, chlorite, actinolite/hornblende, muscovite, etc. with gneissic/schistose texture. Leucosome constitutes light-colored minerals quartz and feldspars. Migmatite indicates the end stage of high-grade metamorphism under water-saturated condition. Based on the combined structure of melanosome and leucosome, migmatite is classified into various types and are briefly mentioned here (Fig. 11.10).

(a) Vein type migmatite: Leucosome forms a network of veins (brownish white) having irregular and thick blocks of melanosome (black colour), enclosing the veins. (Fig. 11.10 a)

Fig. 11.10 (a)

(b) Stromatic migmatite: Leucosome is concordant with melanosome. Melanosome layer is almost parallel to leucosome layer. Both melanosome and leucosome show varying thickness. Sometimes, the leucosomes crosscut melanosome. (Fig. 11.10 b)

Fig. 11.10 (b)

(c) Nebulite: Leucosome forms irregular-shaped patches and grades into melanosomes. (Fig. 11.10 c)

Fig. 11.10 (c)

(d) Agmatite: Numerous blocks of paleosomes are surrounded by narrow or thin veins of leucosome, looking like breccia. (Fig. 11.10 d)

Fig. 11.10 (d)

(e) Rafts are similar to agmatite but having smaller rounded paleosomes and sheared.

(f) Dialation: Stretched component layers are filled by leucosomes. (Fig. 11.10 e)

Fig. 11.10 (e)

(g) Schileran: Stretched, sheared, irrugular shaped melanosome and leucosome. (Fig. 11.10 f)

Fig. 11.10 (f)

11.6.1 Migmatite Genesis

Three theories are put forward to explain migmatite formation.

(a) Injection of granitic material into the dark-colored gneissic/schistose rock.

(b) Localized partial melting. Leucosome is formed by localized partial melting of parent rock gneiss/schist. The unmelted dark-coloured residual rock, called restite, contains garnet, biotite, sillimanite, etc., of the parent rock schist/gneiss.

(c) They are developed by metamorphic differentiation.

Since the chemical and mineralogical composition of paleosome and melanosome seems to be similar to pelitic or parent rock composition, injection migmatite theory is invalid. Migmatite occurs in high-grade metamorphic terrains and appears to be the product of such high grade, where fluid plays a significant role. In 1940 and 1950s, discussion was there to know whether migmatite is magmatic or metamorphic in origin. It is argued that granitic liquid can be generated by water saturated melting of pelites. The composition of leucosome is more tonolitic rather than granitic. Therefore, granitic origin for the formation of migmatite is not favored. The leucosome composition in a single outcrop is not uniform and do not represent the product of single stage partial melting. It is because that there is no uniform melting even in an outcrop scale and there is a possibility of melt loss in melting stage. The metamorphic differentiation is invalid as most of the melting takes place in solid state and very little melt is generated during anatexis. Large-scale melt is not available for differentiation, as in igneous rock crystallization from magma.

11.6.2 Migmatization of High Grade Gneisses by Partial Melting

As local partial melting of the gniesses is more favoured for migmatization, it is better to understand more about the process of partial melting of the gneisses in little more detail. The high grade gneisses generally contain orthoclase, quartz, plagioclase (An_{10-40}), as that of igneous granite, with biotite and minor amounts of sillimanite, cordierite, etc. However, orthoclase and plagioclase proportion vary significantly and the composition of the gneiss may range from granite-tronjhimite-tonolite-granodiorite. Therefore, the gneiss has quartz-orthoclase-plagioclase albite (An_{10-20})-water (SiO_2-$KAlSi_3O_8$-NaAl Si_3O_8-H_2O), and they are represented in Qtz-Or-Ab trainagle, except H_2O, as shown in Fig. 11.11.

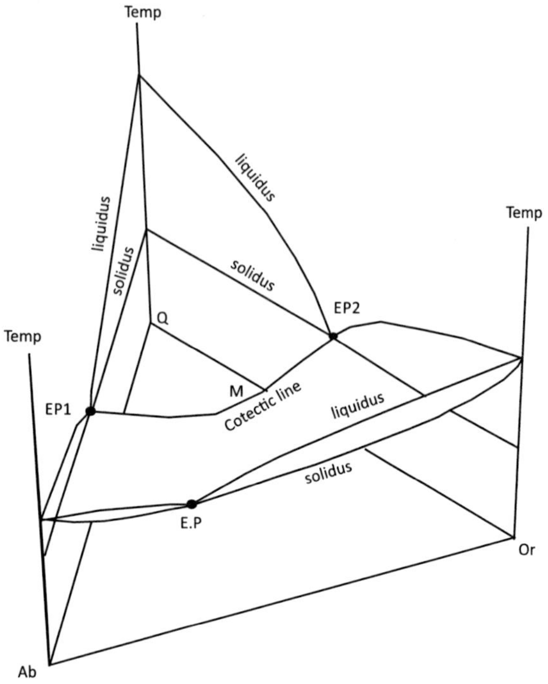

Fig. 11.11: Ternary Diagram for Or-Ab-Qtz System at Constant Pressure of 2.0 - 5.0 kb ($P_{H2O} = P_t$); T – temperature vertical axis. EP – eutectic point; Qtz-Ab EP1 is connected by Qtz-Or EP2 by cotectic line

In the ternary system (Fig. 11.11), Qz-Or, Qtz-Ab and Ab-Or are binary eutectic systems. However, Or-Ab forms limited solid solution in low pressure (<2.2 kb) but above that pressure of >2.2 kb, it becomes eutectic (Fig. 11.11). Two eutectic points EP1 to EP2 is connected by cotectic line. The cotectic line is curved and passed through temperature minimum, shown as M (Fig. 11.12).

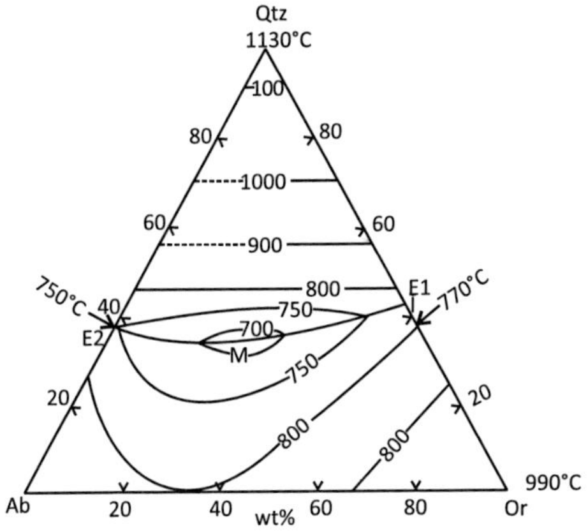

Fig. 11.12: Ternary Diagram Ab-Or-Qtz for Migmatite Melting. Cotectic line E1-E2. M – Temperature minimum, Isotherms are shown as contour lines

Minimum temperature point shifts depending on pressure. If pressure is high, it shifts towards albite side. If pressure is low, it shifts towards orthoclase side. In Fig. 11.12, cotectic line divides the triangulat field into quartz field on the northern side and alkali feldspar field on the southern side. If the overall composition of the gneiss falls in quartz field, quartz melts first and moves towards the cotectic line where feldspar joins melting with quartz. If it falls in alkali feldspar field, feldspar melts first followed by quartz melting along cotectic line.

Many workers carried out melting experiments (see Winkler, 1979 book for details) using paragnesiss with varying ratios of orthoclase-plgioclase-quartz minerals to understand the formation of migmatites. It is observed that paragneiss plagioclase have low An (An_{10-40}). Based on the mineralogy and plagioclase An-content of the paragniess, suitable samples were selected for experiments. In all the experiments, Pt = P_{H_2O} was kept constant at 5.0 kb with increase in temperature.

The results of the experiments are shown in Fig. 11.13 and briefly explained.

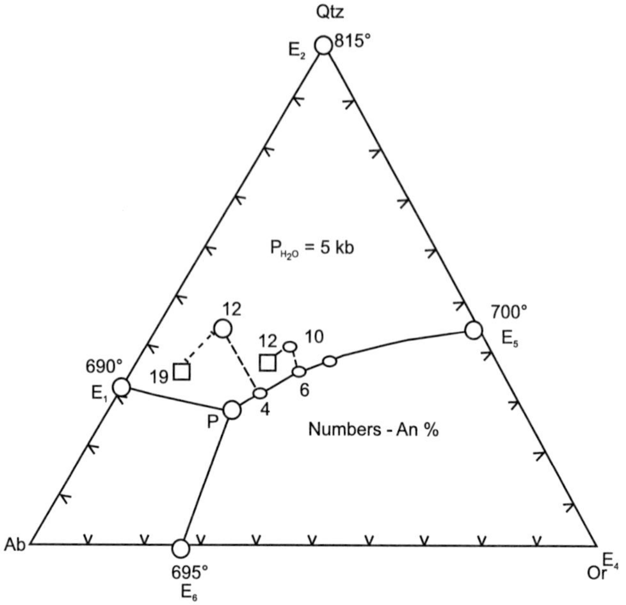

Fig. 11.13: Melting Experiments of Two Paragneiss.
1st sample path 4-6-10-12;
2nd sample path 4-12-19.
E1 – eutectic between Ab-Qtz;
E6 – eutectic between Ab-Or;
E5 – eutectic between Or-Qtz.
P – cotectic point connecting E1-E6-E5

Experiment 1 P = 5 kb

Gneissic sample mineralogy Volume %

Pl – (An_{28})	34
Or	20
Qtz	26
Bt	20

Melting experiment with increase in T

655°C -	solidus
665	All orthoclase melted with residue pl+qtz+bt - 60% melt
675	qtz melted out, residue pl+bt - 75% melt

715	Pl melted out, residue bt - 80% melt

Experiment 2 P = 5 kb

Gneissic sample mineralogy Volume %

Pl (An_{27})	59
Or (Ab_{19})	6
Qtz	26
Bt	9

Melting experiment with increase in T

655°C	solidus
657	orthoclase totally melted out, Residue pl+qtz+bt - 20% melt
700	qtz out, residue pl+bt - 65% melt
740	Pl melted out, residue bt - 75% melt

In both experiments, orthoclase melts first, followed by quartz and finally plagioclase. Biotites do not melt in both the experiments but remain as residue. It may be noticed that in the second experiment, orthoclase melts totally by 2°C temperature increase than first experiment. Plagioclase melted totally in the first experiment than second experiment due to later's higher volume %, even though, An-content in both samples are similar. Biotite is not melted in both the samples, and that has been explained possibly due to higher Mg in biotite.

Another set of experiment with four gneisses of varying plagioclase An-content have been carried out by different workers and their results are given here

Plagioclase	Ab: An ratio	Beginning of melting	Amount of melt at different temperature 700°C	720°C	740°C	770°C
Sample 1	83:17	685°C	48	59	68	73
Sample 2	85:15	685	20	49	69	68
Sample 3	69:31	700	-	31	48	63
Sample 4	62:38	715	-	30	50	70

From the above data, it can be observed that increase in plagioclase An-content delay the melting temperature of the gneiss. Gneiss with lower plagioclase An-content melts at lower temperature than the gneiss having plagioclase with higher An-content. The amount of melt also decreases with higher An-content, e.g. at 740°C, sample 1 shows 68% melting whereas sample 4 shows 50% melting, Only after reaching ~770°C, both samples show nearly equal amount of melting. The above experiments suggest that paragneiss undergo in-situ partial melting in varying scale, due to percentage variation of quartz-K-feldspar-plagioclase minerals as well as plagioclase An-content, to produce migmatites.

11.7 Granulites

As explained earlier, if $P_{H_2O} = P_{total}$ in the almandine amphibolite facies, it leads to melting and migmatisation. However, at the same P-T condition when $P_{H_2O} < P_{total}$, due to the escape of water from the system, a new rock with hypersthene is formed, known as granulite. The rocks are pelite to semi-pelite in composition. Orthopyroxene bearing rock is the type rock of granulite facies metamorphism, irrespective of the rock composition, i.e. pelite, basic or greywacke. It should be termed as granulite only when hypersthene occurs with other minerals in the rock. Any rock occurring within granulite terrain without hypersthene is termed as granoblastite. Such rocks may have clinopyroxene, garnet, quartz, biotite and cordierite minerals. In other words, the rock with hypersthene and other without hypersthene commonly occur together, e.g. Tamil Nadu and Kerala states of south India, granulite and granoblastite rocks are common. Pelitic granulites is leucocratic in colour, fine to medium grained, having >30 to 40% feldspars, ~20% unhydrous Fe-Mg minerals orthopyroxene, clinopyroxene, garnet and quartz. No primary muscovite is associated with quartz and feldspar. The minerals assemblages of these two rocks are given in Fig. 11.14.

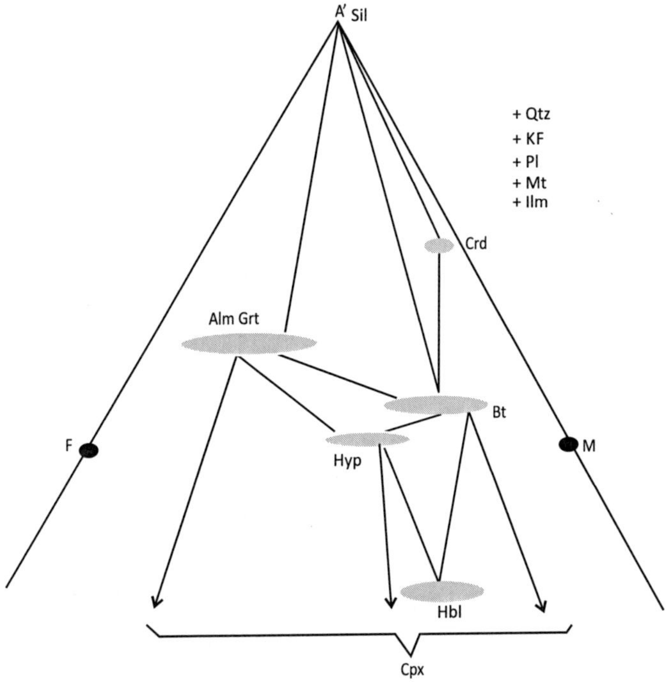

Fig. 11.14: AFM Diagram for Granulite and Granoblastite Rocks —Granulite Assemblage—Alm Grt-Hyp-Bt, Hyp-Alm Grt-Cpx, Bt-Hyp-Hbl; Granoblastite assemblage—Alm Grt-Bt-Sil, Bi-Sil-Crd

11.7.1 Classification of Granulites

It is classified on the basis of quartz-alkali feldspar-plagioclase, like QAP diagram given for igneous rock by Strechisen. Based on the percentage of these minerals, the rock names are given (Fig 11.15). Generally, hypersthene bearing rocks are interlayered with other metamorphic rocks. This classification is only for pelitic to semipelitic composition. The main rock is charnockite, commonly occur in Tamil Nadu state. Charnockite was first dicovererd in Pallavaram hills of Chennai (earlier Madras) by Job Charnock, therefore, the rock name is named as charnockite. This the type locality for charnockite. Based on the percentage variation in K-feldspar, plagioclase and quartz, different rock names are given for the granulites. All these rocks must have hypersthene with QAP minerals to call by different names. These rocks may have hornblende or biotite in minor amounts.

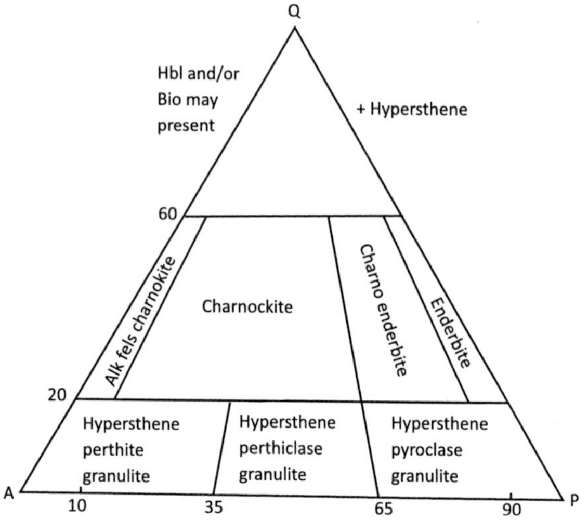

Fig. 11.15: QAP Diagram for the Granulite Classification

Almandine garnet and cordierite occur together in many granulites, but cordierite without garnet is rare (Fig. 11.17a). Biotite decreases in amount with the formation of other Fe-Mg minerals orthopyroxene, clinopyroxene and garnet. Generally, biotite reactions goes to completion in this grade. Sillimanite and hypersthene do not occur together when cordierite and garnet are present in the assemblage (Figs. 11.16, 11.17a). Cordierite, once regarded as low pressure mineral, invariably present with garnet in granulite terrain, even at higher temperature and in certain pressure range. In Al-rich and Mg-rich pelites, the granulites consist of cordierite, biotite and sillimanite, without garnet (Fig. 11.17b).

In some granulites, sapphirine and quartz are present indicating much lower water pressure and higher total pressure of > 8.0 kb. Sapphirine bearing rocks are found in the Kodaikanal hills of Tamil Nadu. Sapphirine formation is explained by the reaction

$$\text{Hypersthene} + \text{sillimanite} \rightarrow \text{sapphirine} + \text{quartz} \qquad (11.25)$$

The assemblage hypersthene, sillimanite and quartz found in some granulite is considered to be equivalent of cordierite high grade, which possibly stabilized at ~650 to 800°C and 10 kb.

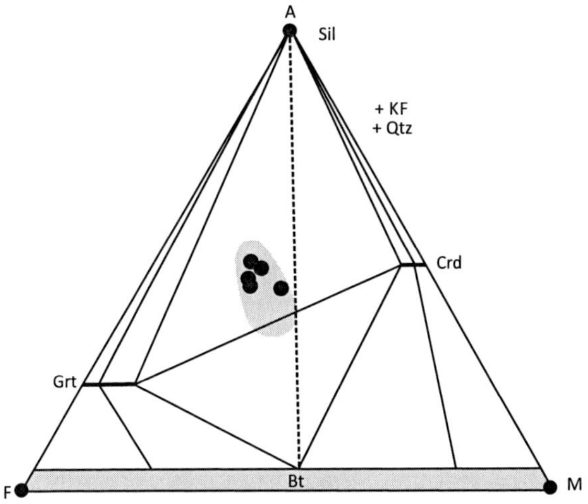

Fig. 11.16: AFM Diagram for Granulite Assemblage Crd-Alm Grt-Sil, Alm Grt-Crd-Bt

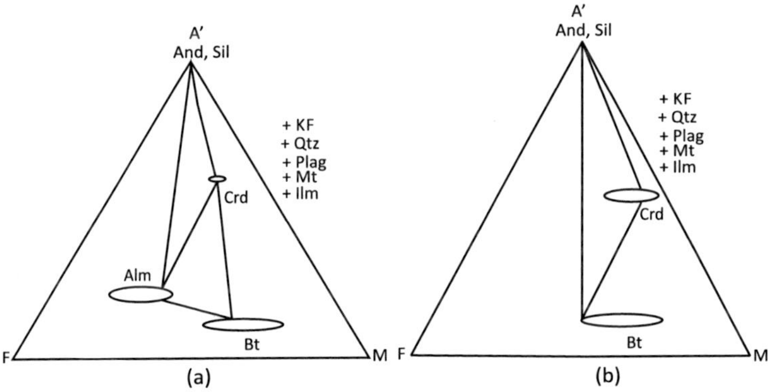

Fig. 11.17: AFM Diagram for Granoblastite (a) Crd-Alm Grt-sil, Alm Grt- Crd-Bt (b) Bt-Crd-Sil Assemblage

11.7.2 Structure and Texture of Granulite

Structure varies from gneissic to massive and granular to granoblastic in nature. Granoblastic equigranular—it is similar to igneous plutonic hipidiomorphic texture. Inequigranular—

it is similar to igneous prophyritic texture of hypabyssal and volcanic rocks. Seriate—the variation in grain size is from fine to coarse grained having gradational texture. Flaser—lenses of strained quartz with undulose extinction are separated by unstarined fine grained matrix minerals. Mylonitic—sheared fine to coarse grained laminated rock.

11.7.3 High Pressure—Low to Medium Temperature Metamorphism

At high pressure and low to medium temperature, metamorphism produces blueschist and eclogite facies rocks. Such type of metamorphism is seen in subduction zones at the accretionary wedge, e.g. Himalayan collision zone (Fig. 3.2). Pelite is characterized by talc, phengite-rich and celedonite-rich muscovite, paragonite, sudite, Fe-Mg poor Al-rich chlorite and carpholite (Fe, Mg, Mn) Al_2SiO_5. Along with phengitic-muscovite, pyrope garnet, chloritoid and kyanite are present in these rocks. Talc is produced by the following reaction:

Mg-Chlorite + phlogopite (Mg-biotite) → Talc + phengite-muscovite + H_2O (11.26)

This reaction is pressure sensitive, in which talc and phengitic-muscovite are stable at a minimum pressure of ~3.0 kbar. In normal pelites, talc is produced above 15.0 kbars in eclogite facies by the reaction

Biotite + chlorite → talc + chloritoid + H_2O (11.27)

The assemblages talc-phengite, talc-kyanite and talc-pyrope garnet are reported in many blueschist and eclogite facies terrenes (Fig. 11.18 a, b). Presence of talc and phengitic-muscovite gives the appearance of white colour, and therefore, called as white schist.

Chlorite disappears producing talc-kyanite assemblage by high pressure reactions. Kyanite is produced in Al-rich pelites at higher pressure by the reaction

Talc + chloritoid → biotite + kyanite + quartz + H_2O (11.28)

Above 18.0 kb pressure, chloritoid disappear producing garnet, biotite and kyanite assemblage. In eclogite facies, coesite (silica polymorph) is reported as inclusion in pyrope garnet with other high pressure mineral assemblage talc, phengitic-muscovite, pyrope garnet and kyanite (Fig. 11.18 c).

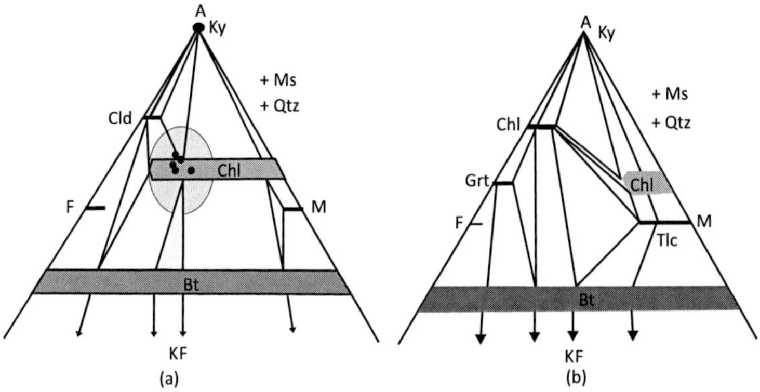

Fig. 11.18: High Pressure Eclogite Facies Assemblage (a) low T eclogite: Bt-Cld-Chl, Cld- Chl-Ky (b) medium T eclogite: Pyp Grt-Cld-Bt, Cld-Chl-Tlc, Cld-Bt-Tlc

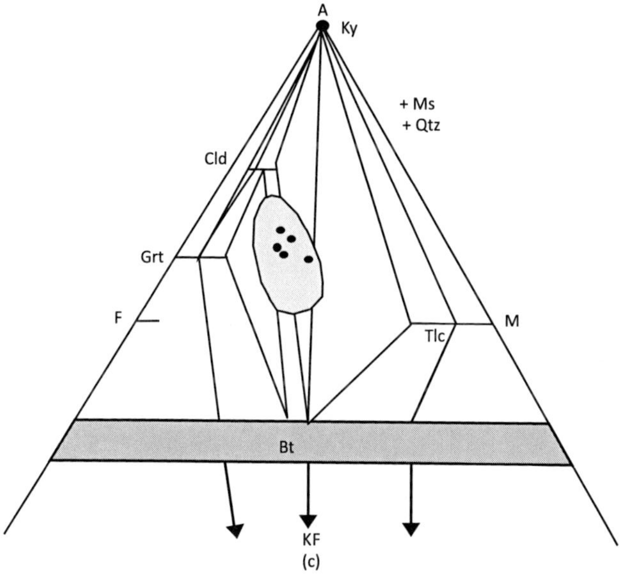

Fig. 11.18: (c) High T eclogite; PypGrt-Cld-Ky, Pyp Grt-Cld-Bt, Bt-Ky-Tlc

11.8 Contact Metamorphism

As mentioned earlier, this metamorphism takes place at shallow depth and is effective with increase in temperature, but with little or no effect with pressure. This is similar to Buchan or Abukuma type regional metamorphism. Pelite shows four metamorphic facies or zones producing metamorphic aureoles. Starting from low to high temperature and low pressure, the four facies are (Fig. 8.1),

(a) Zeolite-prehenite-pumpellyite facies

(b) Albite-epidote hornfels facies

(c) Hornblende hornfels facies

(d) Pyroxene hornfels facies

Shale/slate rocks at low metamorphic grade are called as phyllites or argillites. At low temperature and low to medium pressures, both contact and regional metamorphic mineralogy is very much similar. The rock may have phengite, quartz, chlorite, pyrophyllite, K-feldspar, detrital muscovite and biotite. No mineralogical change is observed in phyllites in zeolite and prehenite-pumpellyite facies.

11.8.1 Albite-Epidote Hornfels Facies

The first mineral to appear is biotite together with muscovite at ~440°C in this facies by reaction

Chlorite + K feldspar → Biotite + muscovite + quartz + H_2O (11.29)

The rock is fairly coarse grained where muscovite and biotite are larger in size and randomly oriented. Chloritoid may appear (Fig. 11.19a) at slightly high temperature in Fe-rich pelites by reaction

Chlorite + pyrophyllite → chloritoid + quartz + H_2O (11.30)

When chloritoid reaches Mg-rich composition, it breaks down to andalusite and chlorite by continuous reaction.

If the composition of pelite is Al-Fe rich as well as in bulk Fe/Mg ratio andalusite is produced at low pressure by reaction

Pyrophyllite → andalusite + quartz + H_2O (11.31)

But this reaction is rarely observed.

At still higher temperature chloritoid and chlorite disappears at ~2.0 kb pressure by the reaction producing biotite and andalusite (Fig. 11.19b).

Chlorite + chloritoid + muscovite → biotite + andalusite + quartz + H_2O (11.32)

Cordierite is formed by the reaction at ~450°C

Mg-Chl + andalusite + quartz → Mg-cordierite + H_2O (11.30)

The main mineralogical change that can be distinguished from high to low pressure, i.e from regional to contact metamorphism, is by the presence of andalusite or cordierite. It is possible to have stable andalusite and cordierite at this grade, but do not occur due to restricted whole rock composition.

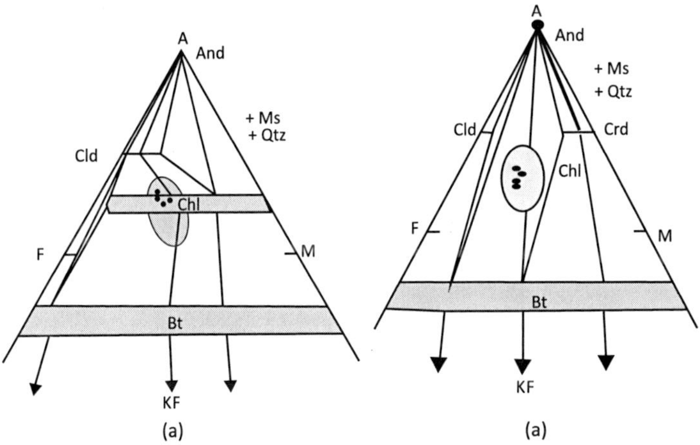

Fig. 11.19: AFM Diagrams for Albite-Epidote Hornfels Facies. (a) Bt-Cld-Chl, Chl-Cld- And (b) And-Bt-Chl, Crd-And-Bt

11.8.2 Hornblende Hornfels Facies

In this facies condition, chlorite is replaced by andalusite-biotite, biotite-andalusite-cordierite and biotite-cordierite assemblages with increasing Mg/Mg+Fe ratio (Fig. 11.20), Cordierite in Mg-rich and andalusite in Al-rich compositions are observed in this grade. If the pressure is > 2.0 kb and at higher temperature, staurolite is formed by the reaction (Fig. 11.18)

Chloritoid + andalusite → staurolite + quartz + H_2O (11.34)

If the pressure lies between 1.0 to 2.0 kb, staurolite is stable at 560°C. If the pressure is < 1.0 kbar, staurolite is unstable.

Garnet may form from chlorite or chloritoid, but it is not stable at low pressure, and therefore, not found in this facies.

11.8.3 Pyroxene Hornfels Facies

In this facies, Fe-rich cordierite or almandine garnet is formed at ~650°C and ~2.0 kb by the reactions (Fig. 11.21)

Biotite + andalusite → cordierite + quartz + muscovite + H_2O or (11.35)

Fe-chloritoid + annite biotite → garnet + quartz + H_2O (11.36)

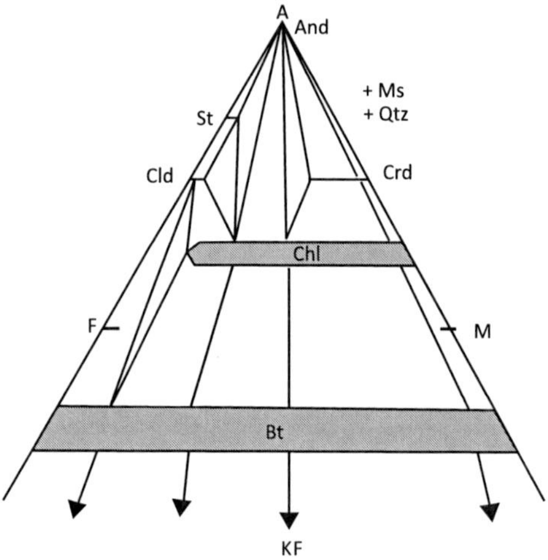

Fig. 11.20: AFM Diagram for Hornblende Hornfels Facies. Cld-St-Chl, St-And-Chl, Chl-Crd-And, Cld-Chl-Bt

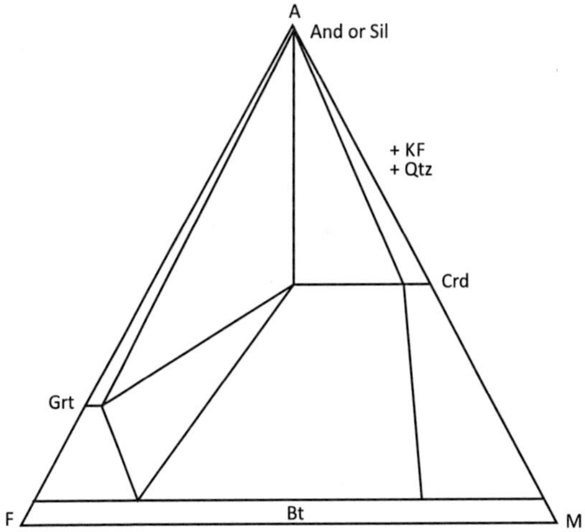

Fig. 11.21: AFM Diagram for Pyroxene Hornfels Facies. Grt-Crd-Sil, Grt-Bt-Crd, Crd- Bt-Qtz

Above this temperature, generation of partial melting is observed. At lower pressure, granitic melting increases with increase in temperature. However, at the above-mentioned temperature and low pressure, instead of partial melting, biotite and muscovite reactions produce orthopyroxene and spinel minerals.

Chapter 12

Metamorphism of Calcareous Rocks

Calcareous rocks include limestone, dolostone and other carbonates that are mostly formed in passive stable continental shelf region. The carbonates range in composition from pure to impure varieties. Impure carbonates may have clastic sediments such as clay, sand, silt, chert, ironoxide, etc. in variable amounts. If clastic sediments and carbonates are in equal amounts, it is known as marl. Likewise, if ironoxide and carbonates are in equal amounts, it is called as iron-rich carbonates. In metamorphic terrains, thin or thick carbonate sediments are interlayered with other sediments like shale, sandstone, greywacke, etc. In case the carbonates have increasing amount of Ca-Mg-Fe silicates like diopside, Ca-amphibole, grossular garnet, epidote, etc., the rock is termed as calc-silicate.

Metamophism of pure limestone, having only calcite, is recrystallised to marble. Marble shows variable grain size depending on metamorphic grade. Low-grade marble is fine grained and high-grade marble is coarse grained. Deformed and sheared marble show preferred orientation of calcite, dolomite, etc., and other associated silicate minerals. Dolostone and calc-silicate rocks give information about de-volatilization and de-carbonation reactions, fluid compositions and P-T condition of metamorphism than pure limestone. The carbonates have CaO, MgO, SiO_2 (CMS) together with volatiles CO_2 and H_2O. These volatiles have variable concentration prior to (pore space H_2O and CO_2) de-volatilization reactions take place. The impure carbonates produce minerals such as talc, tremolite, diopside, fosterite, etc., depending on pore space

$H_2O:CO_2$ ratio and P-T condition of metamorphism. The volatile ratio varies considerably depending on the percentage of carbonate and silicate minerals present at the time of metamorphism. If carbonates are more, CO_2 is released more reducing H_2O content. If more hydroxyl minerals are involved in the reactions, H_2O is more than CO_2. This change in volatiles ratio in impure calcareous rock is due to metamorphic reactions and that is known as buffering. It is discussed in somewhat detail here. The calc-silicate minerals are shown in CMS diagram (Fig. 12.1).

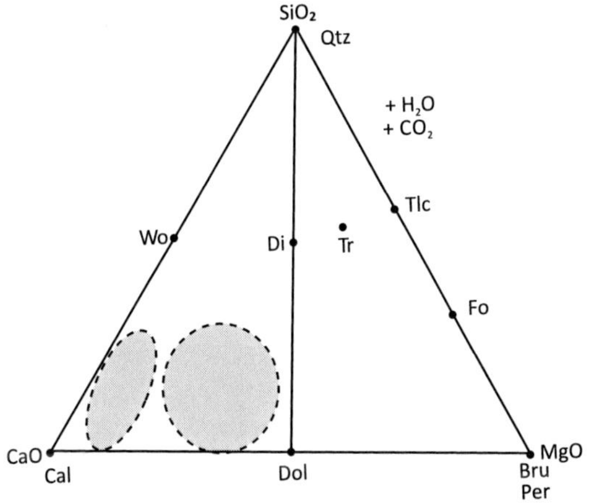

Fig. 12.1: Plot of Calc-Silicate Minerals in CMS Diagram. Big shaded area—average composition of dolomitic limestone; oblate shaded area—average composition of impure limestone

Contact metamorphism takes place at shallow depth (~3 kms or less), with less than 1 kb pressure. Regional metamorphism takes place at deeper level in subduction zones (~15 kms depth or more) at pressures of ~5.0 kb or more. Temperature and fluids play significant role than total pressure in carbonate metamorphism. In view of this, contact and regional metamorphism are discussed using T-X_{CO_2} diagrams at a constant pressure rather than in P-T diagram.

12.1 Contact Metamorphism

The impure dolostones intruded by granite/granodiorite pluton show the formation of any or all of periclase, wollastonite, monticellite, spurrite and merwinite minerals, in the innermost hot contact aureole. Going away from the contact zone, as temperature decreases, fosterite is formed away from contact, further away tremolite, farther away talc and at the farthest end no new mineral. As mentioned earlier, CO_2 and H_2O play significant role in the formation of the above-mentioned minerals. To understand relevant mineral reactions that produce these minerals, TWQ computer program for impure dolostone has been developed and constructed T-X_{CO_2} diagram at 1.0 kb (see Winter, 2012 and others for details). The T-X_{CO_2} diagram diagram at 1.0 kb (Fig. 12.2) suggests that talc, tremolite, fosterite and diopside are formed by several minerals reactions in contact metamorphism.

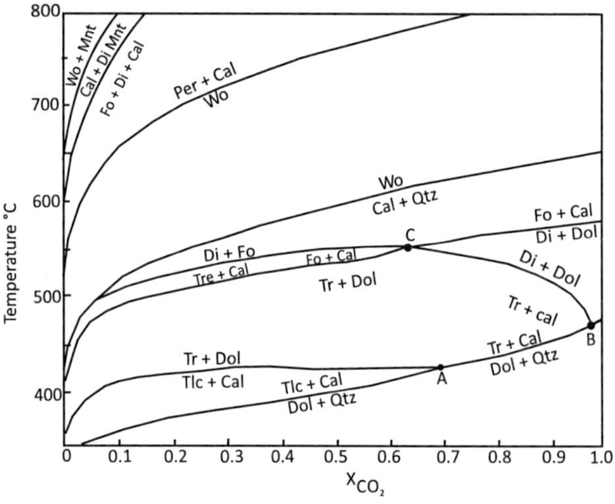

Fig. 12.2: Calcareous Impure Dolostone-T-X_{CO_2} at 1 kb

The reactions are based on two aspects, viz. (a) the percentage of dolomite, calcite and quartz in dolostone and (b) fluids composition X_{CO_2} = 0.1 to 1.0 ($CO_2/CO_2 + H_2O$). Based on T- X_{CO_2} diagram, the following reactions take place that are given below:

(1) No new minerals are formed in the lowest metamorphic grade upto ~200°C and X_{CO_2} = < 0.7.

(2) If the temperature increases from 200°C to ~400°C with < 0.7 X_{CO_2}, following reactions takes place. If the rock has dolomite and quartz, it produces talc by the reaction (Fig. 12.3).

(3) Dolomite+4 quartz+H_2O→talc+3 calcite+3 CO_2 (12.1)

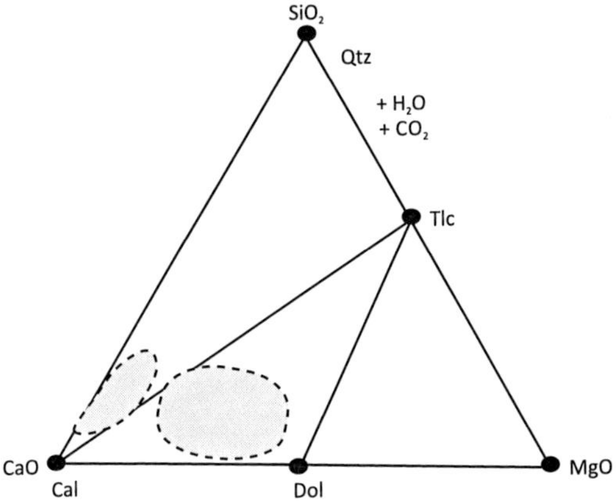

Fig. 12.3: CMS System: Dolomitic Limestone—Equilibrium Assemblage Tlc-Cal-Dol

In case of excess quartz, it combines with talc and calcite producing tremolite by reaction (Figs. 12.2 and 12.4).

5 talc + 4 quartz + 6 calcite → 3 tremolite + 6 CO_2 + 2 H_2O (12.2)

If quartz is less, then (Figs. 12.2 and 12.5) the following reaction takes place

2 Talc +3 calcite → tremolite + dolomite + CO_2 + H_2O (12.3)

If X_{CO_2} = > 0.7 and ~400°C, tremolite is produced instead of talc by the reaction

3 Dolomite + 4 quartz → tremolite + 3 calcite (12.4)

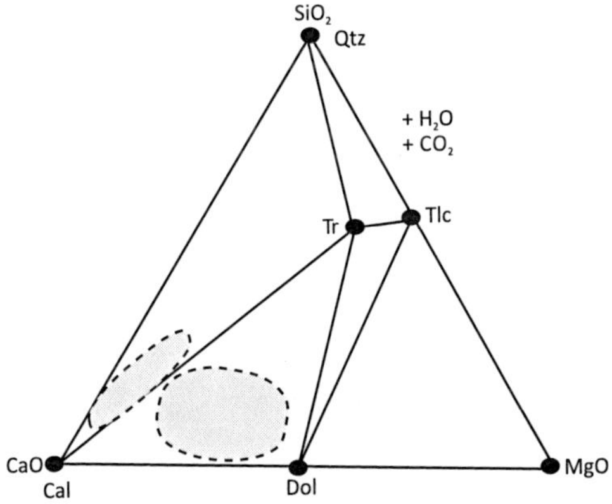

Fig. 12.4: CMS Diagram: Dolomitic Limestone with Excess Quartz—Assemblage Dol-Tre- Tlc; Dol-Tre-Cal

At X_{CO_2} = 0.7 and ~400°C, talc-tremolite-calcite-quartz coexist together forming an invariant point A (Fig. 12.2).

(3) Rising the T to ~450°C with X_{CO_2} = < 0.7, diopside is produced by reaction (Fig. 12.5).

Tremolite + 2 quartz + 3 calcite → 5 Diopside + 3 CO_2 +H_2O (12.5)

This reaction occurs in impure limestone with less quartz.

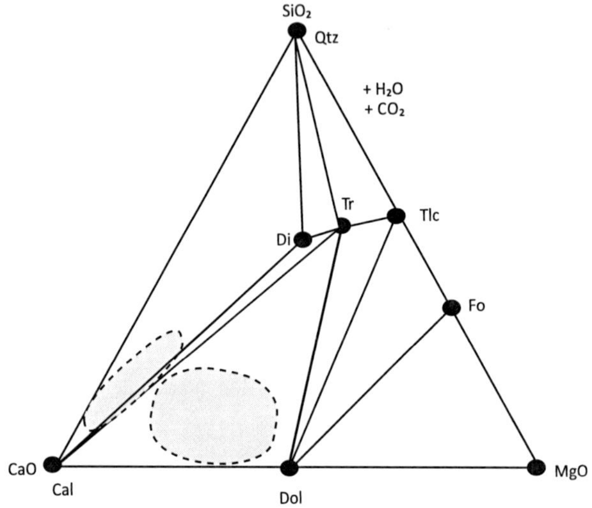

Fig. 12.5: CMS System: Dolomitic Limestone Assemblage—Tlc-Dol-Tr; Di-Cal-Tr

With further slight rise in T and increase in X_{CO_2} = ~0.9, the invariant point B is reached where diopisde–tremolite–calcte – dolomite-quartz assemblage occur (Fig. 12.2).

(4) The increase in T to ~500°C, X_{CO_2} = < 0.63, and without quartz, fosterite and diopside are formed by reaction (Fig. 12.6).

Tremolite + dolomite → 8 fosterite + 13 calcite + 9 CO_2 + H_2O (12.6)

Instead of dolomite, if calcite is present, fosterite and diopside are formed.

With further rise in T to ~550 with X_{CO_2} ~0.7, invariant pint C is reached with tremolite-dolomite-calcite-fosterite-diopside assemblage (Fig. 12.7).

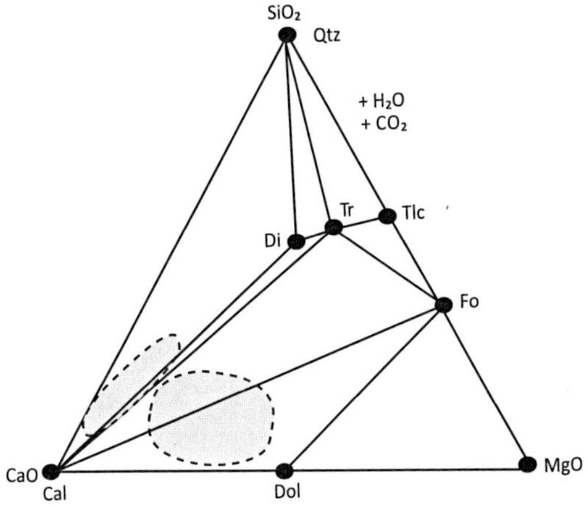

Fig. 12.6: CMS System: Dolomitic Limestone Assemblage—Tre-Fo-Cal; Fo-Cal-Tre

In case of siliceous limestone, at the innermost contact, wollastonite is formed at 500 to 600°C and wide range of X_{CO_2}. > 0.2 to 0.9 by reaction (Fig. 12.7).

Calcite + quartz→wollastonite + CO_2 (12.7)

(5) Above 600 to 700°C, dolomite breaks down to periclase and calcite. Montecellite is formed at ~800°C in high X_{CO_2} and at < 0.2 X_{CO_2}, the same may form at 600 to 650°C.

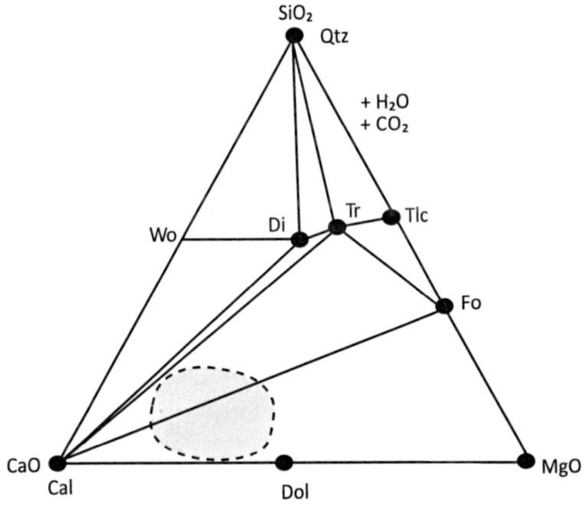

Fig. 12.7: CMS System: Dolomitic Limestone Assemblage—Wo-Di-Cal; Fo-Cal-Tre

The minerals akermannite, spurrite, mervinite and larnite are observed in some impure dolostones, mostly formed at sandinite facies condition.

12.2 Regional Metamorphism

Regional metamorphism takes place at depth and the metamorphic reactions of calcareous rocks produce similar silicate minerals talc, tremolite, diopside, fosterite, etc. as observed in contact metamorphism. Buffering of H_2O and CO_2 is limited, due to closed system behavior, as there is little or no addition of volatiles from external source. H_2O : CO_2 ratio changes during metamorphism depending on initial pore space ratio, and the amount of carbonate/non-carbonate minerals involved in the reactions. Initially, H_2O may be more than CO_2 in pore space fluid. CO_2 may increase more than H_2O by carbonate reactions, and buffering of H_2O may be less due to limited amount of hydroxyl minerals involved in the reactions. Similar to contact metamorphism, T-X_{CO2} diagram at 5.0 kb (Fig. 12.8) has been developed for high pressure to a depth of ~15 kms.

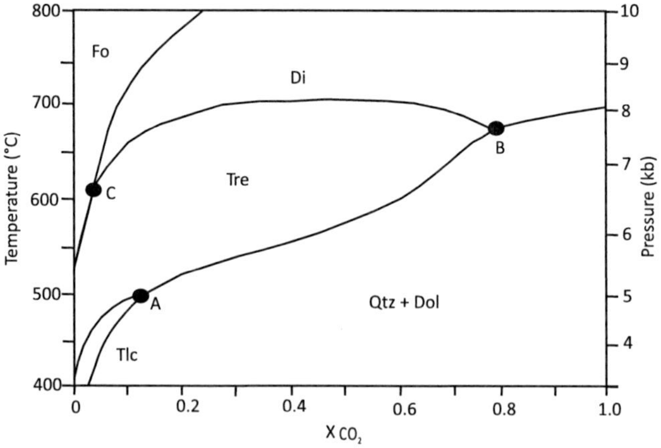

Fig. 12.8: T-X_{CO2} Diagrams for Calcareous Rock at High Pressure for Dolomitic Limestone

The figure suggests two principle effects (a) devolatilisation univariant reactions are curved in shape and (b) reaction temperature increase with increase in pressure. The mineral reactions in contact and regional metamorphism are almost similar, excepting that the reactions in regional metamorphism take place at higher P-T conditions. The invariant points in contact and regional metamorphism occur at different X_{CO_2} values. At higher pressures, talc field becomes smaller and diopside field becomes larger. Talc is formed at 400 to 480°C with

$X_{CO_2} < 0.1$, reaction (12.8).

At 480°C, tremolite-dolomite-calcite-quartz coexist together at an invariant point A (Fig. 12.8).

When $X_{CO_2} > 0.1$ to < 0.9, T = 480 to 600°C, tremolite is formed by reaction (12.4). Further rise in temperature with this range of X_{CO_2}, reaction 12.5 produces diopside. Diopside is also produced by another reaction

Tremolite + calcite → diopside + dolomite + H_2O + CO_2 (12.8)

With further rise in T, fosterite is formed by reaction 12.6 or by the reaction

Diopside + dolomite → Fosterite + calcite + CO_2 (12.9)

However, if quartz is in excess and $X_{CO_2} > 0.90$ and T ~580°C, diopside is produced instead of tremolite by reaction

Dolomite + 2 quartz → diopside + 2 CO_2 (12.10)

At 600°C, dolomite-quartz-diopside coexist forming second invariant point B (Fig. 12.8).

If T is > 700°C and $X_{CO_2} > 0.10$ to 1.0, wollastonite is formed by reaction (12.7).

In regionally metamorphosed terrain, pressure effect is inconsistent but effect of temperature and fluids are significant.

At 5.0 kb, the sequence of minerals include

Talc – 400 to 480°C, $X_{CO_2} < 0.1$

Tremolite – 480°C, invariant point A with $X_{CO_2} > 0.1$ to 0.9

Diopside – 600°C, second invariant point B with $X_{CO_2} > 0.8$ to 1.0

Fosterite 690°C, third invariant point C with $X_{CO_2} > 0.10$ to 1.0

For example, in 1 kb fosterite forms at 530°C, at 5.0 kb at 690°C and 8 kb at 780°C. Similarly, bulk composition plays an important role. Fosterite is favoured in high Mg/Fe and low SiO_2 bulk composition, whereas diopside is in high SiO_2 and low Mg/Fe composition. Wollastonite is formed in upper amphibolite to granulite facies condition.

Therefore, bulk composition and mineralogy of the parent rock is important to understand metamorphism and deciphering metamorphic grade. In the absence of other rock compositions and the presence of only impure dolomitic limestone, the mineral assemblages described so far is used to decipher metamorphic grade qualitatively. This is due to the effect of volatiles playing a significant role in calcareous rock metamorphism.

Chapter 13

Thermodynamic Application

13.1 Introduction

All natural systems have energy. The universe consists of galaxies, each galaxy with numerous stars, each star with number of planets; each planet has rock, water and atmosphere, etc. All members of the whole universe have energy. If one planet is considered as a system, it is a part of one galaxy system, and so on. Any small energy change in a system leads to change in energy of the surrounding systems.

For example, dropping a rock from the height (potential energy) leads to motion and fall to the ground (kinetic energy), i.e. conversion of potential energy (from height) to kinetic energy (to ground). Mixing acid in water, liberates heat or thermal energy due to chemical reaction. Mechanical energy is converted to thermal energy, e.g. hitting the rock with a hammer. The energy is conserved in all process of the system and its immediate surroundings. If we loose energy from a system to its surrounding system, energy is lost in the former and gained by the latter. The rock fall from height to the lowest point, show the higher energy is converted to lowest energy or the stable point. Therefore, the system tends towards minimum energy state which is known as equilibrium state. Sometimes, it may not reach the minimum energy state, but remain in the middle stage for a time being, which is known as metastable or unstable state, and finally goes to minimum energy or equilibrium state.

Thermodynamic application to petrology is well known and applied to understand magmatic crystallization of igneous rocks, mineral reactions in metamorphic rocks, and very low to high temperature crystallization of ore deposits. This happens when there is change in temperature, pressure and composition of the respective systems. The chemical reactions bring changes in thermal or internal energy of the system till minimum energy is reached. Mineral/chemical reactions always take place due to internal energy change, and expressed either in joules or calories (1 calorie = 4.184 joules). 1 calorie is defined as energy given to raise temperature (T) to 1 gram of water to 1°C.

In this chapter, the thermodynamic application is discussed briefly to show its importance in geology. Fundamental aspects of thermodynamics application to geological problems are described in the books by Winter (2012), Wood and Fraser (1976) and R.S. Sharma (2016). As thermodynamics deals mostly with mathematical expression and data, most of the mathematical expressions and thermodynamic data are taken as it is from these books. For more detail information, it is advised to refer these books as well as other advanced books and journals on the application of thermodynamics to geological problems.

13.2 Thermodynamic Laws

First Law. First law states that energy is neither created nor destroyed. When thermal energy transfers heat, the energy lost in one place is gained in another place. Heating the rock leads to chemical reaction where thermal energy of the rock is transferred to another state of the system.

Second Law. Heat energy is transferred from hot to cold bodies, but not in the reverse way. This is given by the expression $dS = dQ/T$.

Third Law. No system reaches $T = 0$ condition. Transfer of heat – given by symbol q, work – by symbol w. By first law – $\Delta E = q - w$ (heat – work).

13.3 Free Energy

Gibbs formulated energy parameter for chemical systems, mainly for rock systems, which is given as

$$G = H - TS \tag{13.1}$$

where H = enthalpy or heat content,

S = entropy or randomness,

T = Temperature

Enthalpy H – heating water to boiling stage, in which hot water has lower heat content, steam has higher heat content

Entropy S – randomness in lowest in crystals or solids Liquid has higher randomness than solids Gas has very high randomness

Energy cannot be defined in terms of absolute quantity. It is defined in terms of relative energy change in mineral reactions or chemical processes. However, energy of a system/phase/mineral is fixed for a given P-T condition. When a system undergoes change of state, G also changes. Initial value of energy (G_1) at a given initial P-T is changed to (G_2) in some other P-T condition.

Difference in energy change is defined as

$$G_2 - G_1 = \Delta G \tag{13.2}$$

G has definite value irrespective of the method the change takes place. Energy is defined in terms of heat and work. Stable form of any chemical system, may be solid/liquid/gas or their combination, means that system has minimum Gibbs energy for a given P, T and X (composition). Gibbs free energy changes when T, P and X is changed. Therefore, using this energy concept, we try to find out whether the system is stable for the given P, T and X of interest. Therefore, to understand chemical reactions and their relation to Gibbs energy for rock systems, certain basics principles are to be known.

In Chapter 7, several mineral reactions are briefed for the formation of metamorphic rocks. Many minerals are formed by reactions that took place due to change in P-T condition. To understand whether the reactions reached equilibrium state or minimum energy state, it is necessary to evaluate the reactions using Gibbs energy. In the following discussion, Gibbs energy and equilibrium condition is applied, initially for a phase (mineral), followed by reaction with more number of pure phases (single component phases) and lastly the reaction with multicomponent phases.

13.4 Thermodynamic Application for Phase

Take a simple phase, mineral quartz (SiO_2). For this mineral or for that matter any mineral, absolute Gibbs free energy cannot be determined. However, it is possible to determine free energy change for the minerals when P-T is changed, provided standard state free energy data is available for the minerals. Quartz free energy at standard state is determined by the following method. Most common reference state for pure elements is 25°C or 298°K and 1 bar pressure.

To get free energy of quartz for this reference/standard state, the following reaction is used

$$Si + O_2 = SiO_2 \tag{13.3}$$

This is written as

$$\Delta G_{f^\circ} = \Delta H^\circ - T\Delta S^\circ \tag{13.4}$$

ΔH of SiO_2 is determined using 1 mole of Si with 1 mole of O_2 by calorimetry. Entropy of any substance at 0°K and 298°K is always zero. From this H and S data, molar Gibbs free energy of formation of quartz is given as

ΔG_{f° of quartz = – 856.3 kj/mole

This data indicates that quartz is stable at 1 bar and 298°K.

G is an extensive variable, as it is dependent on the number of moles present in the phases of the reaction, so it is a molar property. Standard state free energy data for many phases (minerals) are determined by this way. Thermodynamic data of H, S, V for most of the minerals are available from Robie and Waldbaum, 1968, Helgeson et al., 1978 and others to calculate G at standard state and other P-T conditions.

Gibbs free energy change of a phase for any other P-T condition can be determined once reference state data is available. Free energy data of quartz at standard state is already determined. Free energy change of quartz at elevated P-T can now be calculated using the following equation

$$dG = VdP - SdT \tag{13.5}$$

where dG= change in free energy of the phase in other P and T condition

dP = volume change of the phase with respect to P

dT = entropy change of the phase with respect to T

V = volume, S = entropy,

G, V, S are molar property. These parameters are based on number of moles present in any phase or a mineral.

Here, quartz has fixed composition. If there is change in P and T, which is written as dP and dT, G will change as dG. For quartz at elevated P-T, the integration of the equation (13.5) gives

$$G_{P_2T_2} - G_{P_1T_1} = \int_{P_1}^{P_2} VdP - \int_{T_1}^{T_2} SdT$$

$$T_2 = \text{Temp of interest, } T_1 = 298°K \qquad (13.6)$$

$$P_2 = \text{Press of interest } P_1 = 0.1 \text{ MPa}$$

For this integration, it is to necessary to know how V varies with P (isothermal compressibility) and S varies with T. V with P for solids is almost constant for large range of pressures. However, it is not constant for liquids and gases. Similarly, S varies with T for solids, liquids and gases. The relationship is expressed as

$$dS = (Cp/T)\, dT \qquad (13.7)$$

Cp = heat capacity, i.e. heat required to raise 1 mole of substance to 1°C.

If P-T variation is small, it is assumed that V and S are constant for solids. Now, free energy calculation is carried out for quartz at elevated P-T

At standard state $G_{qtz} = -856$ kj/mole.

Now calculation is done for 500 MPa and 500°C for quartz.

To calculate G at 500 MPa, molar V of quartz is required, that is given as

$$V = -22.69 \text{ cm}^3/\text{mole}$$

Now using equation (13.4) $\quad dG = VdP - SdT$

Initial calculation is done for 500 MPa pressure keeping T constant (dT = 0), the equation becomes

$$dG = VdP \tag{13.8}$$

$$G_{P_2} - G_{P_1} = \int_{P_1}^{P_2} VdP = V(P_2 - P_1) \tag{13.9}$$

$P_2 = 500$ MPa; $P_1 = 0\ 1$ MPa at constant T

$G_{0.1,\,298} = -\ 856.300$ j

$G_{500,\,298} = G_{0.1,\,298} + V\ (500 - 0.1)$

$= -\ 856300 + 22.69\ (499.9)$

$= -\ 844957$ J or 845 kj

Next calculation is done for T = (273 + 500) = 773°K at constant P

$$G_{T_2} - G_{T_1} = \int_{P_1}^{P_2} - SdT = -\ S(T_2 - T_1) \tag{13.10}$$

$T_2 = 500°K$; $T_1 = 298°K$ at constant P

$S_{0.1,\,298} = 41.46$ J/k mole

$G_{0.1,\,773} = -\ 856300 - 41.46\ (773 - 298)$

$G\ \widehat{500,\ 298} = -\ 844957 - 41.46(773 - 298)$

$= -\ 844.6$ kj

Using the available thermodynamic data of the phases, it is possible to determine dG of a phase at elevated P-T, to know whether it is stable or unstable in that condition.

13.5 Variation in G for a Reaction with Change in P-T

Most of the reactions in metamorphic rocks are from solid to solid phases and additionally sometimes with gas phases. In igneous rocks, the reactions are mostly from solid to liquid (melt) or vice-versa. In any reaction, if ΔG is negative, the reaction will proceed towards right, as the product has lowest free energy for the given P and T. If ΔG is positive, the reaction will proceed towards left, as the reactant has lowest free energy for that P and T.

In the following discussion, solid to liquid reaction with same chemical composition is taken to understand free energy change with change in P-T condition.

First take only change in T at constant P.

The equation (13.5) dG = VdP – SdT, becomes (dG/dT)P = – S (13.11)

Since S is randomness, with increase in T, the slope for both solid and liquid will be –ve. However, liquid has steeper slope than solid as in Fig. (13.1). At point A solid has lower G, and with increase in T liquid has lower G. However, both phases show decrease in G with increase in T.

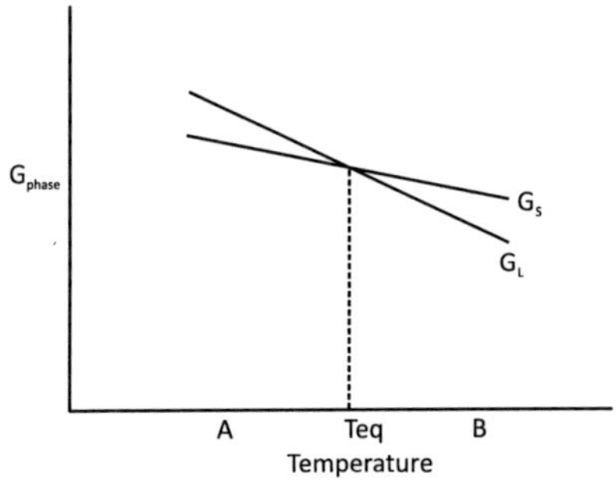

Fig. 13.1: Slope of Reaction Solid to Melt G vs T

Second, take change in P at constant T.

The equation (13.5), SdT, becomes (dG/dP)T = V (13.12)

Here, V liquid > V solid, and slope of liquid > solid

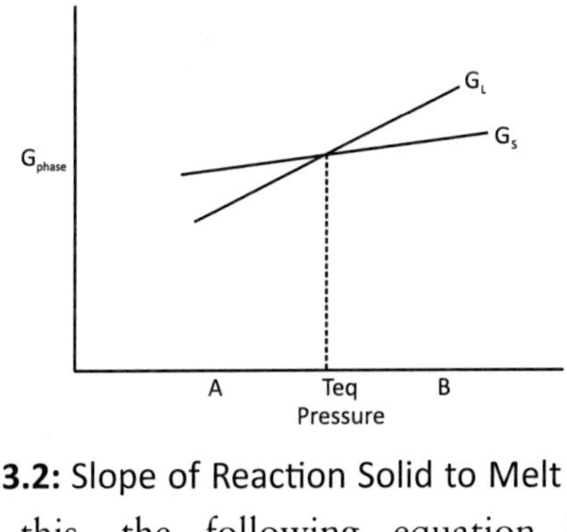

Fig. 13.2: Slope of Reaction Solid to Melt G vs P

Knowing this, the following equation is applied to understand solid to solid reaction,

$$d\Delta G = \Delta V dP - \Delta S dT \tag{13.13}$$

dG represents change in G of a phase with finite change in P and T,

dΔG represents change in ΔG of the reaction with change in P and T

ΔV – volume change between reactant and product

ΔG – free energy difference between reactant to product.

When G of the phase changes with P and T, ΔG of the reaction also changes with P and T. Now the equation (13.13) i.e. $d\Delta G = \Delta VdP - \Delta SdT$ is applied to construct equilibrium curve for the following reaction.

$$\text{Albite} \longrightarrow \text{Jadeite} + \text{Quartz} \qquad (13.14)$$

Ab Jd Qtz

At equilibrium, $\Delta G = 0$, which is valid for a range of P-T and written as

$$\Delta G = \mu_{NaAlSi_2O_6}{}^{Jd} + \mu_{SiO_2}{}^{Qtz} + \mu_{Na\,Al\,Si_3O_8}{}^{Ab}, \qquad (13.15)$$

when they are pure components, it can be written as

$$\Delta G = G^{Jd} + G^{Qtz} - G^{Ab} = 0, \qquad (13.16)$$

Applying equation (13.13), at constant P with variable T, the equation becomes

$$(d\Delta G/dT)_P = -\Delta SdT$$

Likewise at constant T with variable P, it becomes

$$(d\Delta G/dP)_T = \Delta V$$

However, to construct equilibrium curve, the equation is written as

$$d\Delta G = 0 = \Delta VdP - \Delta SdT, \text{ or}$$

$$\Delta VdP = \Delta SdT, \quad \text{it is rewritten as}$$

$$\frac{dP}{dT} = \frac{\Delta S}{\Delta V} \qquad (13.17)$$

This function is known as Claypeyron equation which gives the slope of the equilibrium curve for reaction (13.14) that separates albite from jadiete + quartz.

The thermodynamic data from Wood and Fraser (1976) is taken to construct the equilibrium curve.

	H cal/mole	S (e.u.)
Low Albite	– 937146	50.295
Jadeite	– 719871	31.90
Quartz	– 217650	9.66

ΔG (reaction 13.14) — $G_{Jd} + G_{Qtz} - G_{Ab}$, (13.18)

since G = H–TS, calculation for equation (13.18) is done by

$$\Delta G = 0 = \Delta H_{1,298K} - T\,\Delta S_{1,298K}$$

$$\Delta H_{1,298K} = \Delta\, T\,\Delta S_{1,298K}$$

$$T = \Delta H/\Delta S = -375/-8.42 = 44.5\text{ K}$$

$$\Delta H_{1,298K} = -937146 - (-719871) - (-217650) = -375$$

$$\Delta S_{1,298K} = 50.20 - 31.90 - 9.88 = 8.42$$

$\Delta V = -16.98$ cm^3, which is converted to calorie by dividing with 41.84 = –0.4058 cal/bar^{-1}

Now using equation (13.16),

$dP/dT = dS/dV$; $dP/dT = -8.42/-0.4058 = +20.75$ bar per degree

At 298 K, $P = 1 + 20.75(298 - 44.5)$, where $dP = (P - 1)$ and $dT\ (T - 1)$

= 5260 bar or 5.3 kbar

At 900 K $P = 1 + 20.75(900 - 44.5)$

= 17750 bar or 17.75 kbar

In the same way, if thermodynamic data is available for the metamorphic reactions, it is possible to construct equilibrium curve in the P-T diagram for those reactions.

13.6 Reaction with Solid and Gas/Fluid Phase

When gas or fluid phase is involved in the reaction, effect of pressure on volume and G is different, which is written as

$$d\Delta G = \Delta V dP \qquad \text{or} \qquad (13.19)$$

$$(d\Delta G/dP)_T = \Delta V \text{or}$$

$$\Delta G = V\,(P_2 - P_1)$$

when pressure is applied to fluids, volume changes considerably, unlike it is in solids. To explain this change with respect to P and V for the gas, the ideal gas relationship is applied, which is given as

$PV = nRT$; n = number of moles of gas, $V = RT/P$, (13.20)

R = gas constant i.e. 8.3144 J/mol K.

This relationship is applied and the equation (13.19)

$$d\Delta G = \Delta V dP$$
$$= \Delta(RT/P)/dP,$$

by integration - 1/P dP is ln x, and it becomes

$$= RT \ln P_2 - RT \ln P_1$$
$$= RT \ln (P_2 - P_1)$$
$$= RT \ln (P_2 / P_1)$$

The equation is written as $G_{P,T} = G^0_T + RT \ln P/P_0$. Free energy of gas phase at certain P and T is equal to free energy at some reference state 1 bar and T. T is constant in the calculation but it may be any T.

If the gas behaves non-ideally, P of the gas equation is changed to (f) - the fugacity, $f = P\gamma$; γ = fugacity coefficient, it varies with P and T and also with the proportion of other gas mixtures. In geological problems, H_2O and CO_2 are the main gases involved in most of the reactions, for which γ data is available in literature.

If the reaction involves solids and gases, the equation becomes

$$\Delta G_{P,T} = \Delta G^0_{P,T} + \Delta V_S (P - P^0) + RT \ln (P/ P^0) \qquad (13.21)$$

13.7 Chemical Potential

So far, it is seen that all participating phases in the reaction have fixed composition or pure components and no solid solutions. In nature, most minerals are not pure but are solid solutions and sometimes with mixed gases or fluids. When compositional term is introduced, the number of moles of components i, j, k, etc., to be considered and the primary equation becomes

$$dG = VdP - SdT + \Sigma \mu_i dn_i \qquad (13.22)$$

where μ_i = chemical potential (i.e. free energy G of component i)

n = number of moles of component i in the system

There must be $\mu_i dn_i$ for each component,

μ_i is defined as $\mu_i = (dg/dn_i)$ P, T, n, $j \neq 1$

n, $j \neq 1$ = except i component, all other components are held constant. Example olivine is a solid solution between

Fosterite (Fo) and Fayalite (Fa),

Mg_2SiO_4 and Fe_2SiO_4, where $G^{Fo} = n_{Fo}\ \mu_{Fo}$; $G^{Fa} = n_{Fa}\ \mu_{Fa}$;

G^{Fo} and G^{Fa} = molar free energies of pure fosterite and fayalite

$G_{Total} = n_{Fo}\ \mu_{Fo} + n_{Fa}\ \mu_{Fa}$;

μ of the component of a pure phase (μ^0_{FO}) can be calculated at 1 bar and T of interest from thermodynamic data of H^0 and S^0 (1,298) in the way it is done for

$\Delta G^0 = \Delta H^0_{298} - T\Delta S^0_{298} + (P\text{-}1)\ \Delta V = 0$, where $\Delta G^0 = \Delta\mu^0$, if V is also known, μ^0 can be calculated at any P-T.

13.8 Activity-Composition Relationship

Since this is mostly of mathematical expression, it is taken as it is from R.S. Sharma (2016) book.

Now, look into the activity of component in phase A under specified P-T with reference state, it is given in terms of μ

$\mu_A = \mu^0_i + RT\ \ln a^A_i$, if it is an ideas gas, then activity term becomes $a^A_i = (P_i/P_0)$, and for real gas $a^A_i = = (f_i/f_0)$

The gas in the reference state is

$\mu^A_i = \mu^A_i + RT\ \ln a^A_i = 0$ or $a^A_i = 1$, (13.23)

Rearranging this equation

$$a_i^A = \exp \frac{\mu_A - \mu_i^0}{RT}$$

At standard state 1 bar and T

$\mu^A_i = G_{1,T^0} = H_{1,T^0} - T(S^0)_{1,T}$. At P bar and T

$$\mu_A = \mu_i^0 + \int_1^P \Delta V.dP$$

at standard state 1 bar and T of pure component i in A

$\mu^A_{i\ (P,\ T)} = \mu^0_i$ which at P bar and T

$$\mu^A_{i\ (P,T)} = \mu_i^0 + \int_1^P \Delta V.dP \text{, since } \mu_A - \mu^0_i = RT\ \ln a^A_i$$

$$RT\ \ln a^A_i = \int_1^P \Delta V.dP$$

$RT\ \ln a^A_i = (P-1)\ \Delta V$

If the phase is gas, the volume of the phase has to expressed as a function of P

$PV = RT$ or

$V_A = RT/P$, substituting

$$RT\ \ln a_i^A = \int_1^P RT/P.dP = RT \ln P - RT \ln 1 = RT \ln (P/1)$$

$$a_i^A = P$$

For real gas, it is changed to fugacity (f) $a_i^A = P/P^0$,

$$a_i^A = f/f_i^0,$$

Reaction boundary using solid and fluid phase of the following reaction

$$CaCO_3 + SiO_2 = CaSiO_3 + CO_2 \qquad (13.24)$$

Calcite + quartz = wollastonite + CO_2

At equilibrium

$$\mu_{CaSiO_3}^{Wo} + \mu_{CO_2}^{fl} - \mu_{CaCO_3}^{Cal} - \mu_{SiO_2}^{SiO_2} = \Delta G_{i\ (1,T)}^0 = 0, \qquad (13.25)$$

by substituting

$\mu_i^A = \mu_i^0 + RT \ln a_i^A$ for each component, this becomes

$$\mu_{CaSiO_3}^0 + RTln\ a_{CaSiO_3}^{Wo} + \mu_{Co_2}^0 + RTln\ a_{CO_2}^{fl} - \mu_{CaCO_3}^0 + RTln\ a_{CaCO_3}^{Cal} - \mu_{SiO_2}^0 + RTln\ a_{SiO_2}^{Qtzl} = 0 \qquad (13.26)$$

Rearranging this

$$\mu_{CaSiO_3}^0 + \mu_{CO_2}^0 - \mu_{CaCO_3}^0 - \mu_{SiO_2}^0 = RT \ln \left\{ \frac{a_{CaSiO_3}^{Wo} . a_{CO_2}^{fl}}{a_{CaSiO3}^{Cal} . a_{SiO2}^{Qtzl}} \right\}$$

$$\text{As } \Delta\ \mu^0 = \Delta G^0 = -RT \ln \left\{ \frac{a_{CaSiO_3}^{Wo} . a_{CO_2}^{fl}}{a_{CaCO_3}^{Cal} . a_{SiO_2}^{Qtzl}} \right\} \qquad (13.27)$$

Activity term in paranthesis is K, equilibrium constant.

$RTln\ a_{CaSiO3}^{wo} = (P-1)\ V_{Wo}$; $RTln\ a_{CaCO3}^{Cal} = (P-1)\ V_{Cal}$; $RTln\ a_{SiO2}^{Qtz} = (P-1)\ VQ_{Qtz}$; $a_{CO2}^{fl} = P_{CO2}$

$$\Delta G_{(PT)}^0 = -RTln\ a_{CO2}^{fl} - (P-1)\ \Delta V,$$

where $\Delta V = V_{Wo} - V_{Cal} - V_{Qtz}$

Calculation of reaction boundary for the reaction (13.24)

$$CaCO_3 + SiO_2 = CaSiO_3 + CO_2$$

$\Delta H_{(1,900)}^0 = -19343$ cal

$\Delta S_{(900)} = 35.07$ Cal/K

$\Delta V_{(solids)} = 0.47065$ cal/bar

Substituting into the equation

$\Delta\mu^0 = \Delta G^0 = \Delta H^0_{298} - T\Delta S^0_{298} + (P\text{-}1)\ \Delta V + \text{Rtln}\ P_{CO_2} = 0,$

At 5000 bars

$\mu^0_{reaction} = 19343 - 35.07(T) + (5000 - 1)(-0.47065) + 1.987$ ln $5000 = 0$

$= 19343 - 35.07\ (T) - 2353 + 1.987\ (T)\ 8.517 = 0$

$= 16990 - 18.1\ T = 0$

T = 936 K (663°C)

At 2000 bars T = 922 K (622°C);

at 1000 bars T = 884 K (611°C)

at 1 bar T = 552 K (279°C)

Similarly, it can be calculated for the any such reactions.

Example: Take the following generalised type reaction with solid solution minerals

2A + 3B = C + 4D,

At equilibrium, this reaction is expressed as

$\Delta G = \Delta G^0 + RT \ln K$ where K is written as

$$K = \left\{ \frac{a_c . a_{4D}}{a_{2A} . a_{3B}} \right\} \tag{13.28}$$

We can substitute values for activities for a mineral $a_i = (\gamma_i\ x_i)^Y$

for a gas $a_{gas} = f_{gas}/P^0$ or $a_{gas} = \gamma_{gas}\ P_{gas}$

K_D = distribution of components in phases of the reaction. It is the measured value Xi in the phases of the reaction and is given as

$$K = \left\{ \frac{a_c . a_{4D}}{a_{2A} . a_{3B}} \right\} = \frac{X_c \Upsilon_c . X_{4D}\ \Upsilon_{4D}}{X_{2A}\ \Upsilon_{2A}\ .\ X_{3B}\ \Upsilon_{3B}} = K_D . K_\Upsilon \tag{13.29}$$

The application of K or K_D in the reactions with multicomponent phases can be understood more from Geothermobarometry section.

13.9 Geothermobarometry

Geologists try to estimate P-T condition of metamorphism by many methods. Earlier, pelitic mineral assemblages were used to qualitatively assess the metamorphic condition. Mineral assemblages of these rocks from one terrain to other terrains were compared and assessed the metamorphic grade/zone. Minerals assemblages of these rocks were used to define more number of metamorphic grades/zones. Presence of minerals andalusite and cordierite in pelitic rocks indicated low pressure, whereas presence of jadeite and glaucophane minerals in basic rocks suggested high pressure conditions.

Subsequently, empirical calculations using thermodynamic data for many metamorphic mineral reactions have been carried out to determine P-T condition of formation. Simultaneously, experimental studies have been conducted for many mineral reactions at higher P-T conditions to understand natural metamorphic P-T condition. Another powerful technique of experimental petrology is the calculation of equilibrium pressure and temperature from the measured distribution of elements from co-existing phases. It is now known that coexisting minerals show changes in their chemical composition depending on the P-T condition prevailed at the time of formation. It has been observed that compositions of many coexisting minerals are sensitive to change in P-T condition. P-T condition estimated from the distribution of chemical components in natural coexisting mineral assemblages coupled with empirical/experimental thermodynamic data is known as geothermobarometry. If it is used for the determination of temperature, it is known as geothermometer, and if it is for pressure, it is called as geobarometer.

As explained earlier, P-T estimate is done by using the values of K_D from coexisting mineral phases, where T dependence on K is direct and is given as

$$\Delta G = \Delta G^0 + RT \ln K, \text{ in which}$$

$$\Delta G^0 = -RT \ln K \text{ or } \ln K = -\Delta G^0/RT$$

ΔG^0 can be calculated using 0.1 MPa and 298°K thermodynamic data and then estimate at higher P-T. In the calculation, it is assumed that ΔH, ΔS, ΔV do not vary with change in P-T. It may be mentioned that for more rigorous calculation of ΔG^0, Cp and compressibility data are used to compute ΔH, ΔS, ΔV, for details see Wood and Fraser (1976), Spear (1990) and others.

$\Delta G^0 = \Delta H^0 - T\Delta S + P\Delta V + RT \ln K$, rearranging this

$\ln K = (\Delta H^0/RT) - (\Delta S/R) + (\Delta V/RT)_{dP}$. This is the basic equation for calculating P-T from the measured values of distribution coefficient (K). ΔH, ΔS and ΔV are assumed to be constant while calculating K or K_D.

13.10 Garnet-Biotite Geothermometer

The most common assemblage in pelitic rocks is garnet-biotite together with plagioclase muscovite, quartz in greenschist to almandine amphibolite and even upto granulite facies condition. Therefore, most common garnet-biotite pair is taken for geothermometer calculation. Thermometers generally show large ΔH and small ΔV and the slope of equilibrium curve is steep in P-T space.

Garnet-biotite exchange reaction is written as

$Fe_3Al_2Si_3O_{12}$ + K Mg_3 Al O_{10} $(OH)_2$→
Alm.Garnet Phlogopite

$Mg_3Al_2Si_3O_{12}$ + K Fe_3 Al O_{10} $(OH)_2$ (13.30)
Pyr. Garnet Annite

A.B. Thompson (1976) determined this reaction by empirical method which is given as

$$T(°K) = \frac{2444.39 + 0465\ P\ (kbar)}{3R\ln K + 3.1} \quad (13.31)$$

it is reduced to (without any site mixing)

$$T(°K) = \frac{2740 + 0.0234\ P\ (bar)}{\ln K_D + 1.56} \quad (13.32)$$

Ferry and Spear (1978) carried out experiments at constant pressure at variable temperatures. Their geothermometer is

$$T(°K) = \frac{22454 + 057\ P\ (kbar)}{3\ln K + 4.662}, \text{ where } K - K_D \quad (13.33)$$

$$K_D = \frac{(X_{Fe}^{Grt})^3 \cdot (X_{Mg}^{Bt})^3}{(X_{Mg}^{Grt})^3 \cdot (X_{Fe}^{Bt})^3},$$

$K_D = (X_{Fe}^{Grt})^3 \cdot (X_{Mg}^{Bt})^3/(X_{Mg}^{Grt})^3 \cdot (X_{Fe}^{Bt})^3$,

if the reaction is written with one site mixing

1/3 Fe-biotite + 1/3 Mg-garnet = 1/3 Mg-biotite + 1/3 Fe-garnet

Activity of each component is reduced in the equation, and becomes

3R = 3 X 1.987 = 5.961

(12454/5.961=2089); (0.057/5.961=0.00956); (4.662/5.961 = 0.782)

$$T(°K) = \frac{2089 + 0.00956\ P\ (bar)}{\ln K + 0.782}$$

	Grt	Bio
Mg	1.346	2.988
Fe	4.182	2.070
Ca	0.411	–
Mn	0.061	–

$X_{Mg}^{Grt} = 0.243$; $X_{Mg}^{Bt} = 0.591$; $X_{Fe}^{Grt} = 0.757$; $X_{Fe}^{Bt} = 0.409$

$K_D = 0.222$ ln $K_D = -1.50$ At 5000 bars T = 663 (°K). This reaction is much more sensitive to temperature than pressure.

13.11 Garnet-Aluminosilicate-Quartz-Plagioclase (GASP) Geobarometer

The exchange reactions that show large ΔV and small ΔH are potential geobarometers. Out of many such geobarometers, the assemblage garnet-aluminosilicate-quartz-plagioclase (GASP) in pelitic rocks is taken for pressure determination, as this assemblage is present in upper greenschsit facies (garnet grade) to almandine amphibolite/granulite facies (sillimanite grade) rocks. This barometer is formulated by Newton and Haselton (1976, 1981) using calorimetric data and given empirical geobarometer. The reaction for this assemblage is

$$3Ca\,Al_2\,Sl_2\,O_8 \rightarrow Ca_3\,Al_2\,Sl_3\,O_{12} + 2\,Al_2\,SiO_3 + SiO_2 \quad (13.34)$$

Anorthite	Grossular garnet	kyanite/ sillimanite	Quartz for sillimanite

(13.35)

$$P\ bars = \frac{1 + -1005 + T\,(31.058 + Rt\ln K)}{1.305}$$

$$P\ bars = \frac{1 + -13398 + T\,(36.7 + Rt\ln K)}{1.582} \quad \text{for kyanite (13.36)}$$

Where $\ln K_D = a_{Grs}/a_{An}$

$$a_{Grs} = X_{Grs}.Exp\left\{\frac{[3300 - 1.5T(K)(X_{Py}^2 + X_{Py}.X_{alm} + X_{Py}.X_{spess})]}{RT}\right\}$$

$$A_{Pl} = \frac{X_{An}(1+X)}{4}\exp\frac{(1+X_{An})^2}{RT}(2050+9392)X_{An}$$

This geobarometer has been revised further from experimental data and may be referred for more refined work. Likewise for many thermometers and barometers, revised empirical/experimental data are available, and may be referred the relevant books and journals. Winter (2012) has compiled and given geothermometers and geobarometers for almost all types of metamorphic rocks and can be referred for further details.

Some common geothermometers and geobarometers used for pelitic and basic rocks are listed here.

Pelitic Rocks

Geothermometer

Garnet-Biotite	(Thompson, 1976, Ferry and Spear, 1978)
Garnet-Cordierite	(Thompson, 1976, Holdaway and Lee, 1977, Wells, 1979, Bhattacharya and Sen, 1988)

Geobarometer

Garnet-plagioclase-aluminosilicate-quartz (GASP) (Koziol and Newton, 1976, 1981)

Garnet-cordierite-aluminosilicate-quartz (Aranovich and Podlesskii, 1983)

Garnet-plagioclase-muscobite-biotite (Ghent and Stout, 1981) Cordierite-corundum-spinel-sillimanite (Wells and Richardson, 1979; Harris, 1981)

Basic Rocks

Geothermometer Garnet- Hornblende (Graham and Powell, (1977) Garnet-Olivine (O'Nneil and Wood, (1979)

Orthopyroxene-clinopyroxene (Wood and Banno, 1973, Wells, 1979; Powell, 1978)

Garnet-orthopyroxene (Powell, 1978; Harley, 1984; Lee and Ganguly, 1988; Bhattacharya et al., 1991)

Garnet–clinopyroxene (Raheim and Green, 1974; Dahl, 1980; Ellis and Green, 1979; Wells, 1979; Ganguly and Saxena, 1987)

Olivine-clinopyroxene (Powell and Powell, 1971) Magnetite-Ilmenite (Powell and Powell, 1977)

Geobarometer Garnet-orthopyroxene-plagioclase-quartz Newton and Perkins, 1982; (Newton and Chipere, 1988; Bhattacharya et al., 1991; Holland and Powell, 1990; Berman, 1988; Ekert et al., 1993)

Garnet-clinopyroxene-plagioclase-quartz (Newton and Perkins, 1982; Holland and Powell, 1990; Berman, 1988), Orthopyroxene-olivine-spinel-(Stroh 1976)

For more detail information, it is advised to refer Winter (2012), for maximum number of available geothermobarometers.

Chapter 14

Pressure-Temperature-Time Path (P-T-t path)

In Chapter 8, it was described that metamorphic terrains might have undergone Barrovian metamorphism, Buchan metamorphism, granulite facies metamorphism, bluechist-eclogite facies metamorphism or combination of these metamorphic series. In other words, at what time the high pressure metamorphism is changed to medium pressure metamorphism, or low pressure metamorphism is changed to medium pressure metamorphism, is not known clearly. The process starts with the materials at the surface are subducted or buried underneath with change in P-T condition and metamorphism, followed by uplift and erosion in the final stage. Such a cycle of event is known as P-T-t path. However, the development of a particular metamorphic facies series depends on the kind of subduction the rock undergoes and the rate of uplift and erosion. The metamorphic facies in a terrain indicates the presence of sequence of metamorphic facies and metamorphic field gradient.

P-T-t path is explained in many ways. It is possible to observe partial overprint of one mineral assemblage upon another. Presence of relict minerals may indicate a portion of either prograde or retrograde path or both. It is known that in prograde metamorphism thermodynamic equilibrium is maintained due to which earlier metamorphic path may be erased. In retrograde metamorphism, relict minerals are commonly preserved. Geothermobarometers using core to rim compositions of zoned mineral such as garnet with biotite or with pyroxene or any other coexisting suitable mineral assemblage give information about change in P-T during their growth. Other techniques such as O_2 thermometry, fluid inclusion studies, isotopic cooling ages (Ar^{40}/Ar^{39}), etc., are used to understand uplift rate of the terrain, i.e. reverse model, and not about the forward model related to burial and crustal thickening. P-T-t path is understood using forward heat flow models for various tectonic regimes and evaluate them by comparison with inverse model.

Now, it is understood that regional metamorphism is not due to simple burial, but due to crustal thickening, radioactive heat addition and heat generated at convergent plate boundaries. Several heat models (see Winter, 2012 for details) suggest that heat flow higher than normal continental geotherm is required to produce Barrovian metamorphism.

Figure 14.1 explains P-T-t path for common metamorphic rocks. This is only a schematic diagram and it may vary depending on the rate of thickening, type of deformation, heat transfer, magmatism, erosion, etc.

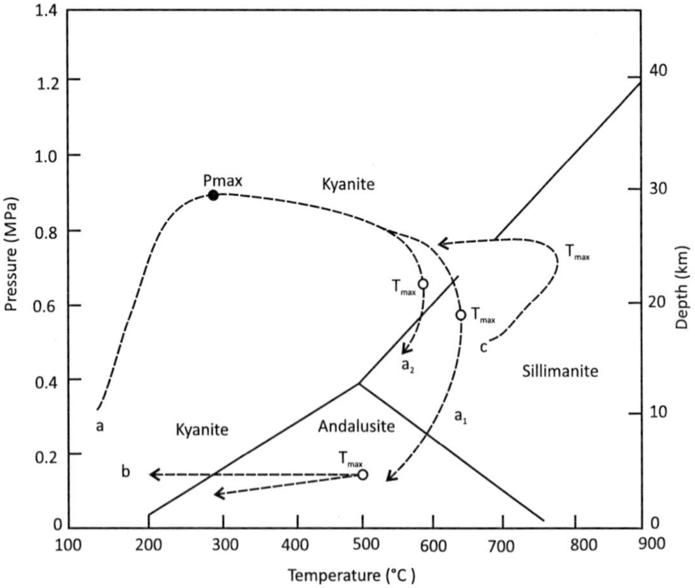

Fig. 14.1: Schematic P-T-t Path for Regional (loop a_1, a_2, c) and Contact (b) Metamorphism

During crustal thickening, pressure increases more rapidly than temperature. To increase the temperature more time is required, because silicate rocks are poor conductor of heat and the rocks get heated very slowly. The thickened crust may reach to depth with quick increase in pressure but relatively remaining cool. The thickened crust is heated by radioactive heat produced by crustal rocks as well as the heat from subduction magmatism. The enhanced heat may remain for a short duration till subduction and thickening of the crust take place. Later uplift and erosion affect the thickened crust and decrease the pressure, before the rocks are equilibrated with higher orogenic geotherm. Temperature still increases due to slow heat transfer, due to which P-T-t path show negative slope after reaching P_{max} (path a_2.) (Fig. 14.1).

The rock reaches T_{max} when uplift and erosion take place with increased geotherm. In the process, thermal perturbation and crustal thickening is dampened or slow down and begin to fade. From this point, P-T-t path follows positive slope and the rocks move to the surface and with decrease in P-T the geotherm returns to normal continental geotherm. When the thickened crust remains stable for a longer time, T_{max} occurs in static elevated geotherm (path a_1) (Fig. 14.1). The geotherm gets relaxed due to erosion during this time. Most of the P-T-t path shows similar kind of loop but the shape of the loop varies depending upon the heat generated by crustal thickening, magmatism, radioactivity, etc. This type of "clockwise P-T-t path" is common in regional metamorphic terrains. It is suggested that positive clockwise loop is mainly due to crustal thickeneing heat flow model.

Anti-clockwise P-T-t path: This path is found only in high grade granulite and gneissic terrains. It is believed that rocks are heated due to mafic magma intrusion at middle to lower crustal level along with truncated or slow uplift and erosion. As the mafic magma is denser, there is slow uplift due to buoyancy. The loop C in Fig. 14.1 suggests T_{max} reaches earlier due to magmatic heating, which is followed by P_{max} showing isobaric cooling. This path indicates uplift or collapse stage after crustal thickening. Both isobaric cooling (path C) and isothermal decompressional (path a_1, a_2) P-T-t path are common in granulite terrains, and the data obtained in such terains should be used with caution.

P-T-t path shown as (b) in Fig. 14.1 indicates different setup where the rock is heated and cooled at constant pressure, due to magmatic intrusion at shallow depth. Such type of P-T-t path indicates contact metamorphism.

However, P-T-t path determined by geothermobarometry might requilibrate the rock that remained at high temperature for extended period of time, and therefore, do not indicate real P-T-t path. The overprinted mineral assemblages also do not really indicate isobaric cooling path. Best way to reconstruct P-T-t path of a metamorphic terrain is done comprehensively by comparing the results obtained from mineral paragenesis, replacement textures, petrogenetic grids and P-T estimates.

Clockwise and anti-clockwise P-T-t path are not restricted only to granulite terrains but also found to occur in medium to high grade metamorphic terrains. In some terrains such as paired metamorphic belts of Japan, Barrovian clockwise P-T-t path and Buchan anti-clockwise P-T-t are observed together. Clockwise P-T-t path is typical of crustal thickening followed by erosion and cooling. Anti clockwise P-T-t path is attributed due to higher heating of the lower crust by numerous plutons derived due to crustal extension and upwelling of the mantle.

In P-T-t paths, P_{max} is reached rapidly with slower erosional unroofing. There is no quantitative estimate of time and the rate of P-T change.

The main points about the P-T-t paths are briefly described here.

(1) Pressure and temperature do not increase uniformely as a single metamorphic grade.

(2) P_{max} and T_{max} do not occur at the same time. In clockwise path, P_{max} is reached much earlier than T_{max}. T_{max} represents higher metamorphic grade at lower pressure than P_{max} pressure. Metamorphic grade is based on mineral assemblages and T_{max} P-T condition.

(3) Temporal mineralogical changes of prograde metamorphism is almost same as in spatial mineralogical changes across the whole terrain.

(4) Many terrains show that spatial association of mineralogical assemblages that follows the field gradient taken along a traverse of a particular terrain.

(5) Clockwise P-T-t path initially follows high pressure blueshist facies series metamorphism and finally equilibrated to greenschsit-amphibolite facies metamorphism at a later time. This pattern of metamorphism is observed in many metamorphic terrains. Such type of metamorphism is found in Alps, Greece, Turkey, and Sanbagawa belt of Japan, northern Asia and China.

(6) If blueschist facies series metamorphic rocks are to remain in a terrain, rapid burial and uplift is required in subduction zones. As most of the uplift in subduction zones is slower, the earlier formed blueschist facies series metamorphism is overprinted by greenschist-amphibolite facies series metamorphism. Presence of bluescshist suggests early stage subduction zone metamorphism.

(7) Recent advancement in textural geochronology (age dating) using small accessory minerals like monazite, zircon, etc., in the rocks gained much importance. The age dating is related to the growth of the mineral with time, get information about specific tectonic-metamorphic events, and enable to quantify the time.

(8) Dating by Sm/Nd, U/Pb and Rb/Sr ages and P-T estimates of zoned garnet has been done for Barrovian zone metamorphic rocks by some workers which is given in Fig. 14.2. The average age of heating is calculated as 8.6°C/Ma and burial at 0.8 km/Ma. The data indicate that from burial point a to reach P_{max}, it took 16. Ma. Rb/Sr dating of coexisting silicate minerals give another 37 Ma to reach T_{max} from which cooling rate is estimated at 4°C/Ma. This suggests that cooling and uplift rate is much slower than heating and burial.

(9) Dating monazite inclusion in garnet and P-T estimates of zoned garnet from the Pakistan Himalayan samples indicate 550°C at 82 Ma and 620°C at 59 Ma, which

gives the heating rate of 2.4 ± 1.2°C/Ma and show clockwise P-T-t path. The age data show monazite growth started at 82 Ma, its growth continued during exhumation at 59 Ma (Fig. 14.2) and melting during subsequent decompression.

(10) Zoned monazite, grown in three to four stages, is dated using U/Pb ages for Canadian Cordillera (Fig. 14.2). The age data is used to constrain the time of thrusting and exhumation. Heating rate is calculated as 5 ± 2°C/Ma.

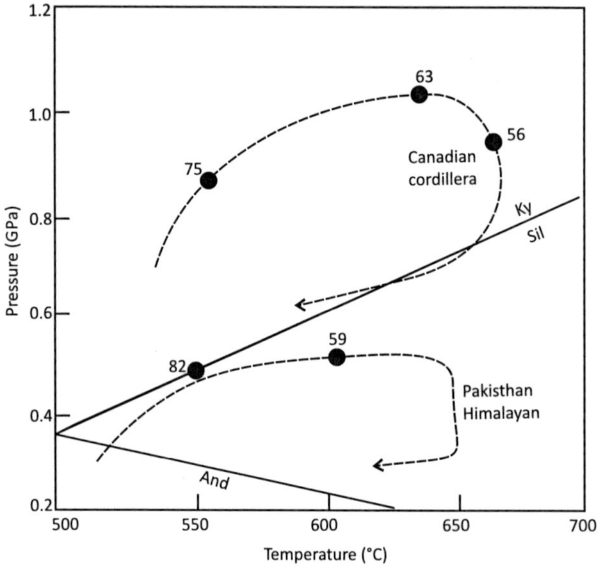

Fig. 14.2

Both clockwise and anti-clockwise P-T-t paths are determined by combining mineral paragenesis, petrogenetic grids, geothermobarometsrs and geochronlogical methods using appropriate mineral assemblages.

Chapter 15

Nomenclature of Metamorphic Rocks

In metamorphic rocks, certain names are common to most or all the metamorphic rock compositions, but for some, the names are specific to a particular rock composition. The names used in metamorphic rocks are described briefly here.

Common sedimentary rocks that are affected by metamorphism include shale/argillaceous or the pelites, weathered basic volcanic/plutonic rocks or sedimentary greywacke's, impure dolostone/limestone, limestone, sandstone and weathered ultramafic rocks.

When shale is metamorphosed, at the beginning stage of compaction or diagenesis shale is converted to slate at about 50 to 150°C. When metamorphic temperature reaches to ~200°C, slate is converted to phyllite. Further increase in temperature, phyllite is changed to mica schist or different types of schists, and at higher grade, it is changed from schist to different types of gneisses. In the highest metamorphic grade, gneiss undergoes melting and produce migmatite or without melting the granulite.

Weathered basic rocks at the beginning of very low grade metamorphism of 100 to 200°C produce fine grained zeolites, prehenite-pumpellyite schists. At low temperature, chlorite schist is formed. At medium to higher temperatures, greenschist is changed to amphibolite. There is no melting at the highest temperatures, but changed to granulites. At low temperatures and higher pressures, the rock is changed to blueschist or glaucophane schist. At still higher pressures and low to medium temperatures, the rock is changed to eclogite. Sandstone contains mainly quartz, and therefore, it is recrystallized to quartzite. Similarly, limestone contains only calcite and dolomite, and is recrystallized to marble, irrespective of metamorphic grade.

Impure limestone, besides calcite and dolomite, contains shale, silica, ferruginous impurities. Depending on the composition of impurities, the impure limestone/dolostone is converted to talc schist in low to medium grade and at higher grades, talc schist is converted to varieties of tremolite schist. Ultramafic rock is converted to low temperature metamorphic rock, due to fluid activity, into serpentinite.

The rock names associated with rock compositions, their mineral assemblages at different facies/grades are given in Table 15.1. The bold and underline names of minerals in each rock composition and in each facies are characteristic of that particular facies.

Table 15.1: Metamorphic Facies/Grades of Basic, Pelitic, Calcareous and Ultramafic Rock Compositions – their names and diagnostic mineral assemblages

Facies	Basic Rocks/greywackes	Pelites/argillaceous	Calcareous	Ultramafic
	Metamorphic names and mineral assemblages	Metamorphic names and mineral assemblages		
Contact Metamorphic facies				
Albite-epidote hornfels facies	Albite- epidote hornfels **Albite + epidote + actionlite** + chlorite	Pyrophyllite-andalusite hornfels **Pyrophyllite + Andalusite + chlorite** + Muscovite + biotite + quartz	Tremolite-calcite hornfels **Tremolite** + calcite + quartz + **epidote + chlorite**	Chl – Tremolite hornfels **Chlorite + tremolite** + quartz + **epidote**
Hornblende hornfels facies	Hornblende hornfels **Hornblende + plagioclase + Diopside** + biotite + quartz	Cordierite-andalusite-muscovite hornfels **Cordierite + Andalusite**+ muscovite + quartz + chlorite + biotite	Diopside-Plagioclase hornfels **Diopside** + **plagioclase** + **grosslar garnet** + chlorite + quartz	Talc-Tremolite hornfels **Talc + tremolite + anthophyllite** + cordierite + biotite
Pyroxene hornfels facies	Cpx-Opx-Hbl hornfels **Clinopyroxene** + **Orthopyroxene** + plagioclase + olivine + **hornblende**	Cordierite-andalusite-K-feldspar hornfels **Cordierite + andalusite / sillimanite** + biotite + opx + quartz K-feldspar	Same as in Hornblende hornfels	Same as in Hornblende hornfels
Sanidinite facies	Anorthite + **diopside + hypersthene**	**Mullite** + Cordierite + **Tridymite + spinel (Magnetite) Corundum** + glass	Same as in Hornblende hornfels	In Mg-rich rocks **Fosterite** + sphene + **periclase** + cordierite

Regional Metamorphic Facies				
Zeolite facies	Zeolites **Laumonite (most typical), analcite, heulandite, wairakite**. Incompletely reacted relics of primary minerals widespread	Slate **Mixed layered/clays Illite crystallinity** changes with increase in T	Tremolite-chlorite schist **Tremolite + calcite** + quartz + **epidote + chlorite**	No clear data
Prehenite-pumpellyite facies	Prehenite-pumpellyite schist **Prehnite + pumpellyite + chlorite** + albite + epidote (lower temperature zone) **Pumpellyite + actionolite** (higher temperature zone) **Lawsonite + albite** (higher pressure zone)	Phyllite **Illite / sericite** + Albite + quartz ± **pyrophyllite** + muscovite In Fe-rich composition **Stilpnomalene** + chlorite + **actinolite** muscovite + albite + quartz	Similar to zeolite facies	No clear data
Greenschist facies	Chlorite Zone Chlorite schist Epidote + **albite + chlorite + stilpnomelane** (lower temperature zone)	Chlorite Zone Sericite schist /phyllite **Chlorite + muscovite/sericite + albite** + quartz (lowest temperature zone)	Tremolite schist **Tremolite + calcite** + albite + quartz	Talc-tremolite schist Talc-**tremolite/actinolite** + quartz + albite
	Biotite Zone Greenschist **Actionolite + epidote + albite + chlorite + stilpnomelane** (lower temperature zone same as above	Biotite Zone Biotite schist **Chlorite + biotite** + albite + quartz + **slipnomalene**	Tremolite schist **Tremolite + calcite** + albite + quartz + epidote	Same as in chlorite zone

	Garnet Zone Greenschist **Hornblende + actionoline + albite** + chlorite + epidote + garnet (higher temperature zone)	Garnet Zone garnetiferrous Muscovite-biotite schist **Garnet + chlorite** + muscovite + **biotite** + albite ± **Chloritoid ± paragonite** (higher temperature zone)	Tremolite schist **Tremolite/actinolite + calcite** + albite + quartz	Greenschist-calc silicate **Hornblende/actinolite + talc + almandine garnet** + chlorite
Lower Amphibolite facies	Amphibolite **Hornblende + plagioclase** + garnet + epidote +	Staurolite-kyanite schist **Staurolite, kyanite** + muscovite + garnet + biotite ± **Mr-rich chlorite + oligoclase plagioclase** (lower temperature zone) + ilmenite	Tremolite-garnet schist **Tremolite + grossular garnet + diopside** + calcite + epidote + plagioclase	Calc schist **Anthophyllite/ cummningte + almandine garnet** + tremolite
Upper amphibolite facies	Almandine Amphibolite facies Same as in lower amphibolite facies	K-feldspar-sillimanite gneiss **Sillimanite + cordierite or garnet + K-feldspar** + biotite + oligoclase + ilmanite	Calc schist **Diopside + hornblende + calcite** + plagioclase	Calc schist **Anthophyllite/ cummningtonite** + tremolite hornblende

Granulite facies	Low pressure Basic Granulite **Orthopyroxene + clinopyroxene + plagioclase + olivine** + hornblende (low pressure) Medium pressure Basic Granulite **Garnet + clinopyroxene + orthopyroxene** + plagioclase + hornblende (medium pressure) High pressure Basic Granulite Garnet + clinopyroxene + quartz + plagioclase + hornblende (high pressure) No melting is generally observed	Medium pressure Acidic Granulite **Cordierite + garnet + K-feldspar + hypersthene** plagioclase + **sillimanite** + quartz (medium pressure) High pressure Acidic Granulite If **kyanite with K-feldspar + garnet + hypersthene** + quartz (high pressure) High temperature Acidic Granulite **Hypersthene + sapphirine + K-feldspar** + quartz (very high temperature) **Migmatite** With partial melting, formation of migmatite, instead of granulite	Calc schist **Diopside + garnet** + calcite + **wollastonite** + plagioclase No other changes No other changes	Ultra Basic granulite **Diopside + orthopyroxene** + hornblende No other changes No other changes
Blueschist facies (Low T and high P)	Blueschist **Glaucophane + lawsonite** + epidote	Phengite-chlorite schist **Phengite + chlorite or talc** + garnet, no biotite, ± **Mg-chloritoid ± carpholite**	Lawsonite-pumpellyite schist **Arogonite + lawsonite + pumpellyite** + calcite	Same as seen blueschsit of basic rocks
Eclogite facies (medium to high T and high P)	Eclogite **Omphacite + garnet**, no plagioclase, no lawsonite	Kyanite-garnet-talc schist **Talc + kyanite + garnet + muscovite (phengitic)**	Similar to eclogite facies assemblage	Same as in eclogite of basic rocks

15.1 Slate. Shale is converted to slate in very low grade metamorphism. It has mainly illite clay, sericite, quartz and chlorite minerals. It shows platy structure.

15.2 Phyllite. It is a fine grained rock produced from shale in low grade metamorphism. It contains platy minerals sericite, phengite muscovite, pyrophyllite, chlorite together with quartz and albite. The picture shows crenulation cleavage that is developed due to the presence of platy minerals.

15.3 Garnet mica schist. This rock contains platy minerals chlorite, muscovite and biotite together with garnet, albite and quartz minerals. It is having schistose structure.

15.4 Biotite muscovite schist. It is another variety of pelitic schist formed from shale rock. It consists mainly muscovite, biotite, feldspars and quartz.

15.5 Kyanite schist. It contains muscovite, biotite, bladed blue-coloured kyanite, plagioclase, quartz and some staurolite. It is showing schistose structure.

15.6 Staurolite in mica schist. It is formed in staurolite-kyanite grade of amphibolite facies metamorphism. Staurolite is generally associated with kyanite, muscovite, biotite, garnet plagioclase and quartz. In the picture, kyanite is seen as light bluish coloured plate with two sets of cleavage.

15.7 Marble. Metamorphic rock made of calcite and/or dolomite derived from sedimentary limestone or dolostone. It is fine to coarse grained depending on grade of metamorphism. Low grade is fine grained and high grade is coarse grained. Marble outcrop (Bluish white colour) is shown here.

15.8 Soapstone/Talc schist. Impure dolostone is converted to talc schist at low grade to medium grade metamorphism. It consists of talc, calcite, dolomite ± chlorite and ± quartz.

15.9 Tremolite schist. Impure dolostone is converted to talc-tremolite schist at medium to high grade metamorphism. It consists mainly of talc, tremolite, ± chlorite, ± quartz, calcite and dolomite. Tremolite occurs in the form of radiating needles.

15.10 Actinolite Schsit. It is formed in garnet grade of greenschist facies metamorphism, derived from basic rock. It contains mainly radiating needles of actinolite together with chlorite and quartz.

15.11 Calc-silicate. Rock containing various types of Ca-Al-Mg-Fe silicates such as grossularite, epidote, tremolite, talc, vesuvianite, etc., and derived from impure limestone or dolostone. These minerals may develop from upper greenschist to upper amphibolite facies metamorphism.

15.12 Skarn. Calc silicate rock formed during metasomatism of limestone country rock by acidic igneous intrusion, where both matter and volatiles are exchanged between both intrusion and the country rock. The minerals include epidote, grossularite, tremolite, etc.

15.13 Quartzite. It is derived from sedimentary sandstone and contains mainly quartz, feldspar and minor amount of silicate minerals. Both brownish and green coloured quartzites are shown here.

15.14 Serpentinite. It is derived from ultramafic rocks peridotite, pyroxenite and dunite, that is produced by low grade metamorphism.

15.15 Augen gneiss. If the porphyroblasts, generally the feldspars in the gneiss, is augen shapped (eye shapped), it is called as augen gneiss. In figure, augen feldspar is wrapped by sheared mica foliation.

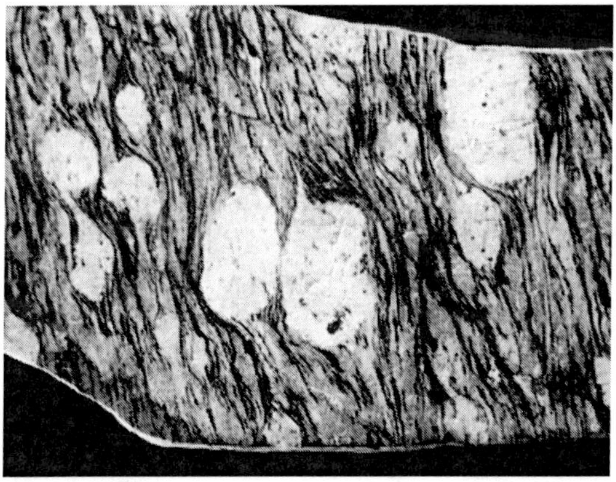

15.16 Migmatite. Peilitic rocks undergo partial melting when the rock reaches almandine amphibolite facies metamorphism. It shows quartz-feldspathic layer alternate with biotite mica or Fe-Mg rich mafic minerals layer. Quartz-feldspar layer is light coloured, it is termed as leucosome. The Fe-Mg minerals layer is dark coloured and is called as melanosome. Generally, the leucosome layers are thicker than melanosome layers.

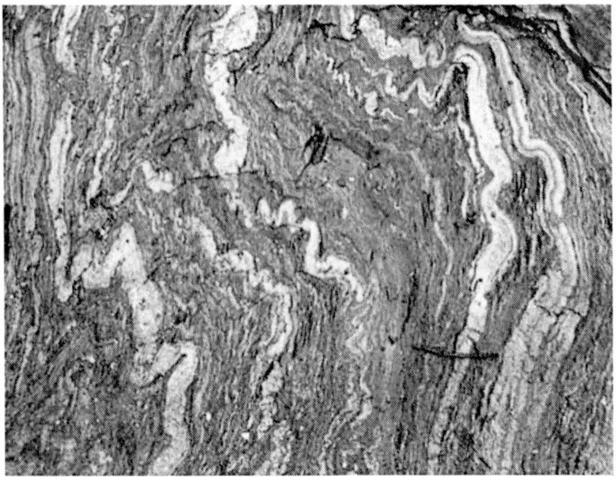

15.17 Porphyroblastic. If the metamorphic rock contains numerous porphyroblasts (bigger crystals) in the fine-grained matrix, it is termed as porphyroclastic.

15.18 Ortho-gneiss. If the metamorphic gneiss rock is derived from igneous rock, it is termed as ortho-gneiss, e.g granite is converted to granite gneiss.

15.19 Para-gneiss. If the gneiss is derived from sedimentary rock, it is termed as para-gneiss e.g. quartzo-feldspathic sandstone is converted to para-gniees, due to metamorphism.

Gneissic rock in the field. Light and dark-coloured minerals in alternate layers.

15.20 Gneiss rock. Light and dark coloured minerals in alternate layers. Light coloured minerals are quartz and feldspars, dark-coloured minerals are biotite.

15.21 Leptinite. Light coloured rock consists mainly of quartz and feldspar, without mica, but sometimes with garnet, formed in granulite facies metamorphic condition.

15.22 Khondolite. Light coloured rock consists mainly of quartz, feldspar, graphite and sometimes with garnet, but without mica, but formed in granulite facies.

15.23 Garnetiferrous amphibolite. It is metamorphosed basic rock. It contains mainly hornblende (greenish black), garnet, plagioclase and quartz (while colour), sometimes with epidote and/or zoisite. It is showing schistose structure in which garnet porphyroblasts (brownish red colour) are present. Second picture is in enlarged scale showing garnet porphyroblast.

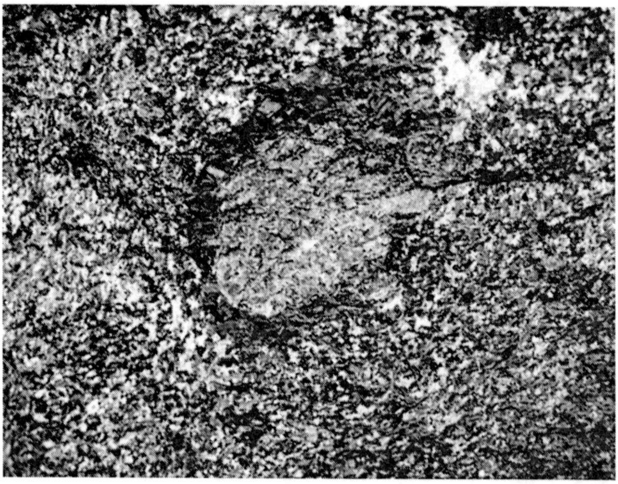

15.24 Amphibolite. It is metamorphosed basic rock. It contains hornblende (greenish black), garnet, plagioclase and quartz (while colour). It is coarse grained and showing granular texture.

15.25 Chlorite schist. It is a metamorphic rock derived from weathered basic rock or greywacke. It contains chlorite, albite, epidote and quartz showing schistose structure.

15.26 Chlorite schist with idioblastic garnet. It is a metamorphic rock derived from weathered basic rock or greywacke. It contains chlorite, actinolite, albite, epidote and quartz showing schistose structure.

15.27 Migmatite. It showing thick dark-coloured bands alternating with thin light coloured bands showing partial melting and folded structure.

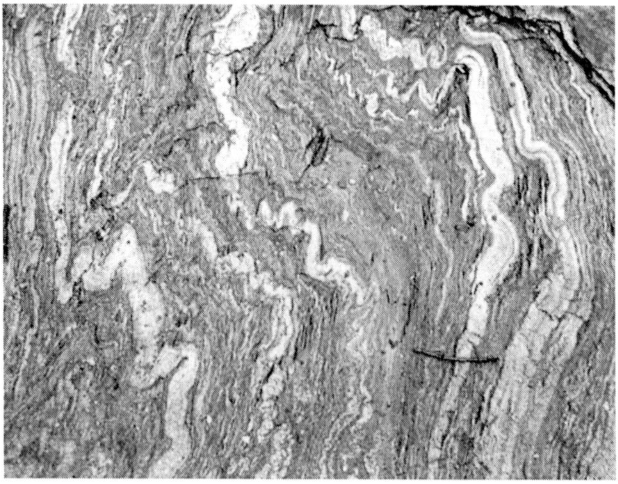

15.28 Typical charnockite. It showing quartz, K-feldsarp, plagioclase and orthopyroxene hypersthenes. Equigranular granoblastic polygonal texture.

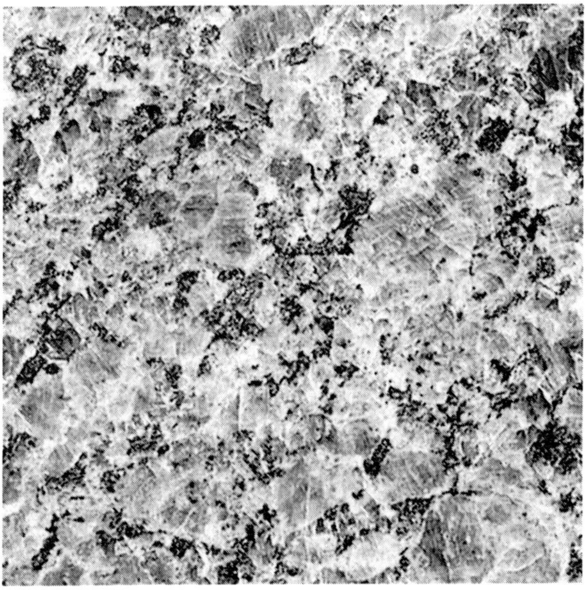

Typical charnockite. Showing quartz, K-feldsarp, plagioclase and orthopyroxene hypersthene. Equigranular granoblastic polygonal texture.

15.29 Garnetiferous granulite (felsic). It is formed in quartzo-feldspathic gneiss or in psammitic rock compositions, under granulite facies metamorphism. It consists of plagioclase, K-feldspar, garnet, quartz, and biotite and with or without orthopyroxene.

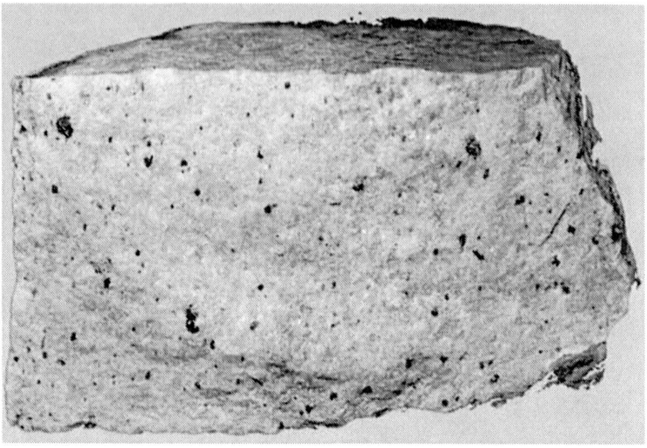

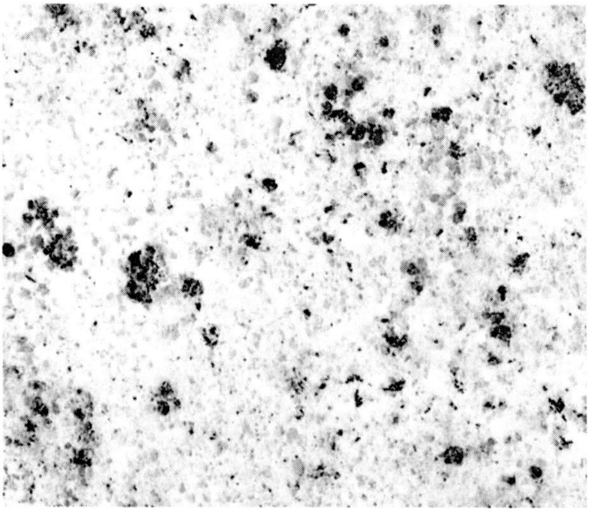

15.30 Garnetiferous granulite (mafic). It is formed in basic rock compositions of granulite facies metamorphism. It consists of plagioclase, K-feldspar, garnet, quartz, orthopyroxene, clinopyroxnene and biotite.

15.31 Garnetiferous 2 pyroxene granulite. Garnet porphyroblast occur along with orthopyroxene-clinopyroxene-plagioclase minerals.

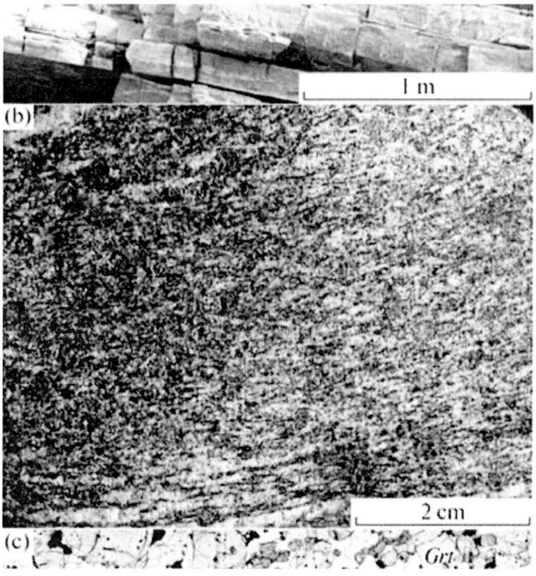

15.32 Sapphirine granulite. It is formed in rocks of pelitic to basic composition. It consists of sapphirine, garnet, plagioclase, orthopyroxene and biotite. It is formed in very high pressure and high temperature condition.

15.33 Blue schist. It is formed in subduction zone in low temperature and high pressure condition, formed from volcanic rock of basic composition. It has glaucophane, lawsonite, garnet and epidote minerals. The rock is blue in colour due to the presence of glaucophane. First figure is an outcrop and second figure is hand specimen.

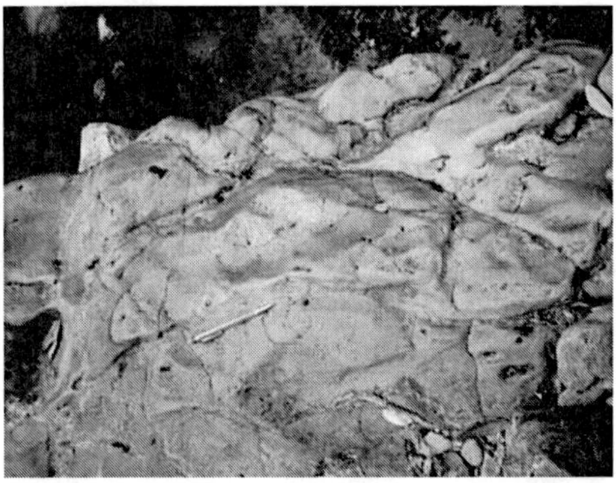

15.34 Eclogite. This rock is formed in volcanic rock of basic composition. It contains omphacite (green colour), garnet (red), kyanite (light blue) and quartz. It is developed in medium to high temperature and high pressure condition. Garnet usually contains inclusions of high pressure quartz coesite.

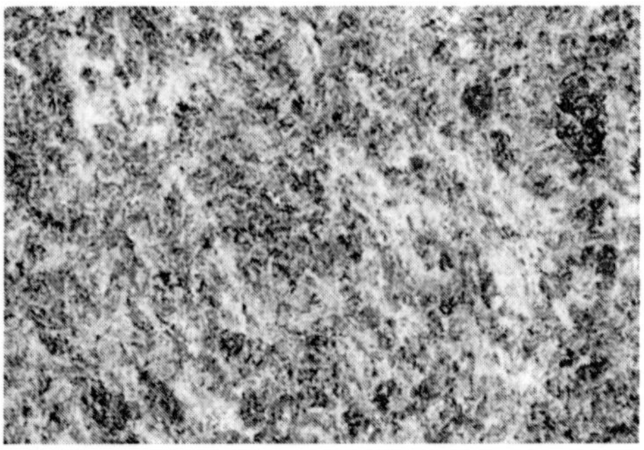

15.35 Blue schist and eclogite rock. Fine-scale layers of eclogite and blueschist showing evidence for ductile deformation.

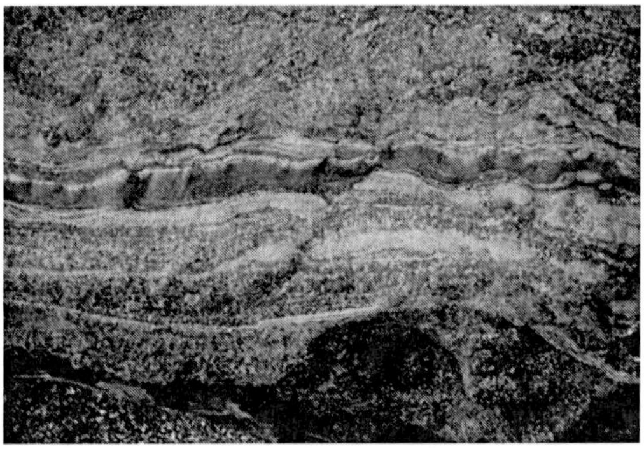

15.36 Peridotite with veins. Peridotite is showing hydrothermal alteration along the fractures with the development of low temperature amphibole.

15.37 Crystotile asbestos (amphibole) in ultramafic rock. The ultramafic peridotite (black colour) is showing crysotile asbestos formation along the fractures in the form of veins.

15.38 Sepentine in ultramafic rock. The ultramafic peridotite (greenish black colour) is showing serpentine (brownish white fibre) formation along the fractures due to hydrothermal alteration.

Chapter 16

Metamorphic Textures

Metamorphic textures or the micro-fabrics are formed in contact and regional metamorphic conditions. Other types of metamorphism, described in Chapter 3, are confined to smaller area, and therefore, not described. Textures in metamorphic rocks are mostly understood from thin section studies. Field and hand specimen study give information about rock type, grade of metamorphism and large-scale intensity of deformation. Regional metamorphic rocks are formed by tectonic processes that took place in subduction/collision zones, inducing changes in mineral assemblages and structure/texture. The rocks in such zones are affected by low to high intensity shearing and might have undergone more than one episode of metamorphism and deformation. The structural features observed in field give information about various stages of deformation. To know the tectonic history of the metamorphic terrain, it is necessary to study rock thin sections for petrography and texture. Based on several studies from the literature, important structures/textures are briefly described here.

Textural Nomenclature in Metamorphic Rocks

The term blast is used as prefix or suffix for porphyritic and poikilitic minerals in metamorphic rocks. If the porphyritic or poikilitic mineral is present in the parent rock before metamorphism, and survived during metamorphism, it is termed as blasto-porphyritic or blasto-ophitic texture. If the porphyritic minerals are formed during metamorphism, such minerals are termed as porphyroblast or poikilioblast. The term relict is used in the rocks, as some part of the original texture is survived during metamorphism and deformation.

If the mineral is euhedral in shape as in igneous rock, it is called as idioblastic. If it has no proper shape or anhedral, it is known as xenoblatic. If it is neither idioblastic nor xenoblastic or but subhedral in shape, it is sub-idioblastic. However, mineral shapes depend on their crystallographic form, temperature, pressure and shearing stress.

16.1 Contact Metamorphic Textures

Contact metamorphic texture is developed in the country rock due to igneous intrusion that takes place at shallow depth where total pressure and deviatoric stress are negligible. The new minerals are formed in a static environment and do not show any preferred orientation. Minerals such as quartz, calcite, etc. are equi-dimensional, whereas mica, amphibole, andalusite, etc., are elongated and occur at random orientations (Fig. 16.1).

decussate

Fig. 16.1: Contact Metamorphic Texture. Crystallised in Static Environment

Relict textures are common in such rocks as there is no deformation. The metamorphic minerals, developed in the country rock, occur as spots of various sizes at random orientations as shown in Fig. 16.2, called as hornfels. Monomineralic rock with perfect grain boundaries is termed as granoblastic polygonal (Fig. 16.2 A) and the texture is termed as granofels. Sometimes two or more minerals have different sizes and polygonal shapes, as in Fig. 16.2 B, C, are termed as granoblastic polygonal.

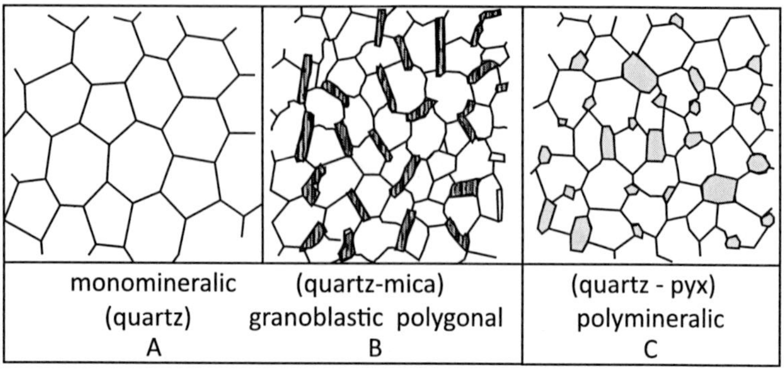

Fig. 16.2: Contact Metamorphic Texture

16.2 Regional Metamorphic Textures

Regional metamorphic terrains show evidences of many metamorphic cycles and several deformational episodes. Thrusting and faulting in the subduction zones produce various deformation textures with or without the formation of new minerals. In shallow thrust or fault zones, the minerals are broken/crushed and recrystallized to various textures. Mechanical fragmentation with sliding and rotation produce fault gauge or fault breccia or cataclasite (Fig. 16.3).

Fig. 16.3: Cataclastic Texture—Feldspar porphyroblast show undulose extension with crushed margins, quartz grains are crushed and recrystallized, micaceous minerals wrapping around porphyroblasts—X nicols

In this process, sometimes, high strained grains show dissolution at their contacts and recrystallize the dissolved materials away from the contact zone, due to the presence of intragranular fluids.

When temperature-pressure increases at depth, the rocks show evidences of metamorphism as well as deformation. During deformation the rock behaves more of brittle-ductile to ductile (elastic) manner. The minerals are bent or elongated due to the slip taking place along point lattice or line lattice or plane lattice. If plane lattice of a mineral slips successively, it is called as dislocation glide and the mineral becomes ductile. If there is dislocation of lattice points in certain part of the mineral, the lattice points gets reoriented and produce undulose extinction as in quartz, mica, etc. The rocks show the development of planar texture the foliation, linear texture the lineation and lattice/dimensional preferred orientations.

16.3 Foliation

Any planer fabric element such as sedimentary banding, arrangement of micaceous minerals in a plane, etc. are called as foliation.

When there is thrusting, right hand side block either moves upward or remain rigid and the left hand side block moves downward creating the thrust effect. For instance, oceanic crust (left hand side block) is subducting beneath continental crust (right hand side block), in which the main movement is of oceanic crust rather than continental crust. Oceanic sediments are affected by thrusting and metamorphism. Thrusting at shallow depth in the subduction zone, the sediment minerals are crushed, pulvarised and recrystallised producing very few metamorphic minerals. However, at greater depth new metamorphic minerals are formed and behave more elastically due to higher P-T condition. In this process, platy minerals mica, chlorite, elongate minerals amphibole, kyanite, staurolite, etc., formed during metamorphism are oriented along the thrust direction producing main foliation, defined as S_0 (Fig. 16.4 c and d).

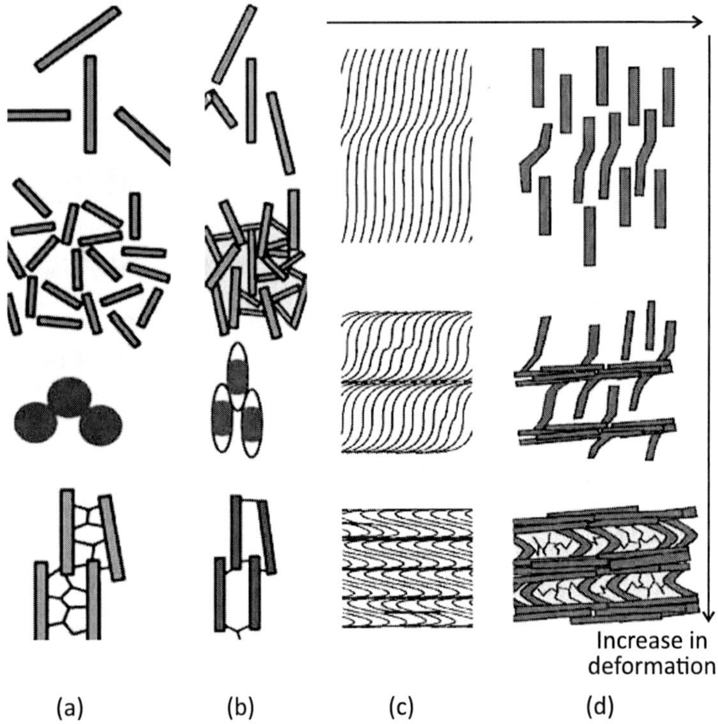

Fig. 16.4: (a) random orientation of minerals in the rock (b) minerals are trying to align or orient along the direction of thrusting, (c, d) Top—primary foliation S_0 showing slight deformation to produce second foliation, bottom 2nd and 3rd—with increase in temperature and continued deformation—production of second foliation S_1

If the primary foliation S_0 is affected by deformation with increase in temperature-pressure, second foliation S_2 is developed as shown in Fig. 16.5. Likewise, if the region is affected by many deformation events, the rock develop more number of foliations, defined as S_1, S_2, S_3 and so on (Fig. 16.6). The metamorphic minerals may be formed (a) prior to deformation, (b) simultaneously with deformation and (c) after deformation. Keeping this view, regional metamorphic textures are defined as pre-tectonic, syn-tectonic and post-tectonic textures. Sometimes, a mineral may start growing prior to deformation, continue to grow during deformation and still grow further even after the deformation is ceased, e.g. garnet porphyroblast grow before deformation, continue during deformation and still grow even after deformation is over. With this view, some of the important regional metamorphic textures are described here. For detailed study on structure and texture of metamorphic rocks, it is suggested to refer books written by Spry, Winter and others.

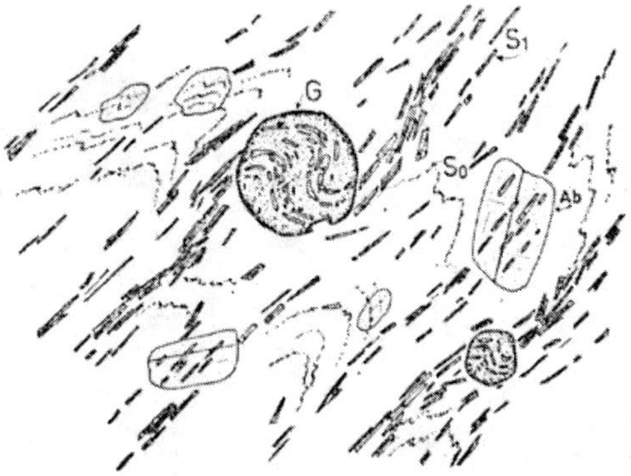

Fig. 16.5: Figure Showing Garnet Mica Schist—albite porphyroblasts show inclusion trails following S_0 (top left); garnet and albite inclusions show their continuation of S_1 foliation (center garnet and right middle albite). S_0 occurs as folded remnants within S_1 foliations

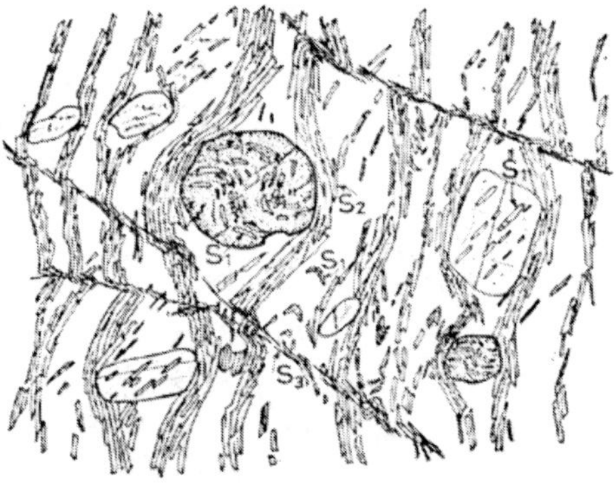

Fig. 16.6: Figure Shows Garnet Mica Schist—garnet and albite porphyroblasts show inclusion trails. S_1 occurrs as inclusions in garnet and albite—cross cutting S_2 main foliation, and occurs within S_2 foliation. S_3 foliation cross cut both S_1 and S_2 foliations

16.4 Lineation

Any linear or oriented or needle shape fabric in the rocks is known as lineation. Figure 16.7a slate rock shows micaceous foliation as numeraous wrinkled fold axes as lineation and (b) biotite mica and feldspar minerals show fibrolite sillimanite needles as mineral lineations. Sillimanite needles are fibrous and black in colour. North-western side feldspars show silighly thick sillimanite needles.

(a)

(b)

Fig. 16.7: (a) Slate Showing Crenulation Cleavage (fold axes) (b) fibrolite sillimanite needles in biotite and feldspar (photomicrograph)

16.5 Pre-tectonic Textures

Regional metamorphism produces new minerals before the beginning of shearing or deformation. Subsequent to minerals formation, the rocks are deformed showing different pre-tectonic textures (see Table 16.1, Section IV, 9).

16.6 Syntectonic Textures

Such textures are developed during metamorphism accompanied by deformation. The new metamorphic minerals and deformation take place simultaneously. In Fig. 16.8a, b, the main foliation continues within the garnet porphyroblast as inclusion trails due to shearing and metamorphism.

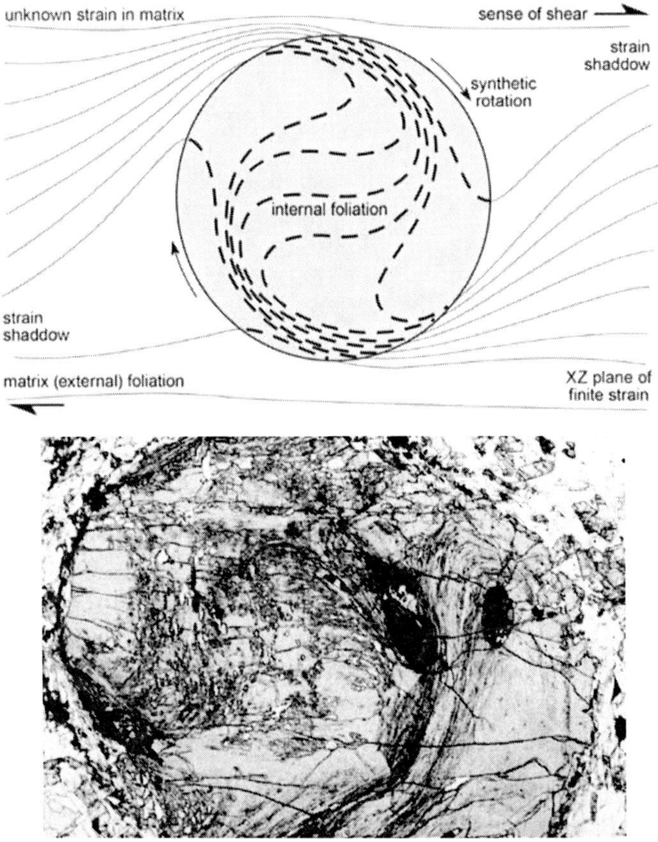

Fig. 16.8: Syntectonic Texture (a) schematic garnet porphyroblast with inclusion trains showing S type inclusion rotation which is continuing with outer main folition, (b) same in actual rock (photopmicrograph)

16.7 Post-tectonic Textures

This kind of texture is developed in rocks where deformation (thrusting) is ceased, but metamorphism is continued. In other words, metamorphic minerals are formed even after deformation. Such textures are termed as post-tectonic textures (Fig. 16.9).

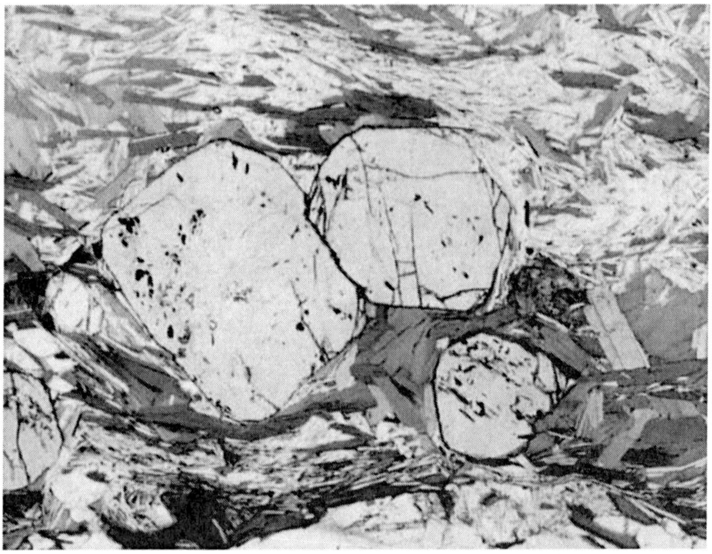

Fig. 16.9: Idioblastic Garnet Prophyroblast Overgrowing Mica Foliation, Suggesting Post-tectonic Growth (photomicrograph)

In the tectonically active zones, metamorphism may start in pre-tectonic condition with the formation of pre-tectonic textures. This may be followed by metamorphism and deformation producing syn-tectonic textures. Finally, with the continuation of metamorphism but with no deformation produce post-tectonic textures (Figs. 16.5 and 16.6).

The description of other textures is given in the following Table 16.1 with suitable figures, wherever necessary.

Table 16.1: Textures of Metamorphic Rocks

I General textural terms		
1	Porphyroblastic	Larger porphyroblats are embedded in fine grained matrix, e.g. garnet, staurolite, cordierite porphyroblasts in fine grained micaceous foliated matrix 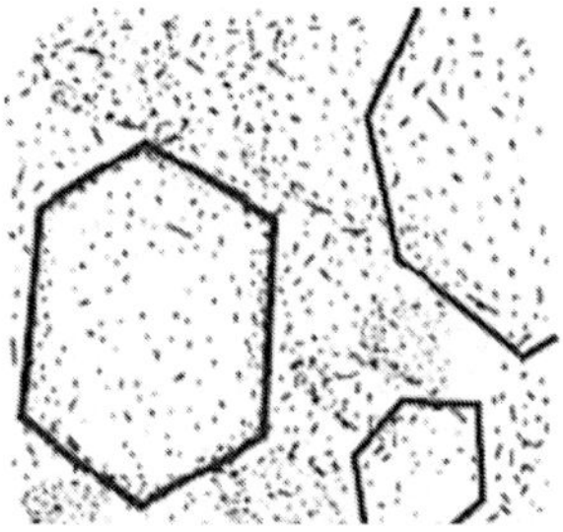
2	Poikilioblast	Schistose rocks may have large poikilioblasts like feldspar, staurolite, cordierite, etc. with numerous mineral inclusions 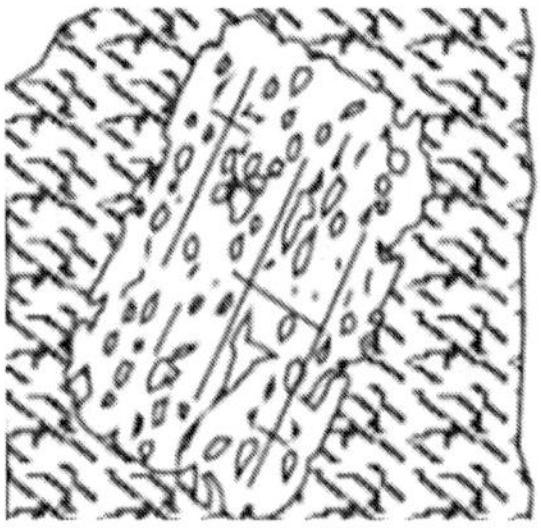
3	Spotted	Small elongated needless of andalusite (white) in fine grained slate matrix (greyish black) that gives the appearance of randomly oriented spots—field photo 

4	Nodular	Oval shaped porphyroblasts such as quartz and rounded rocks embedded in fine grained matrix 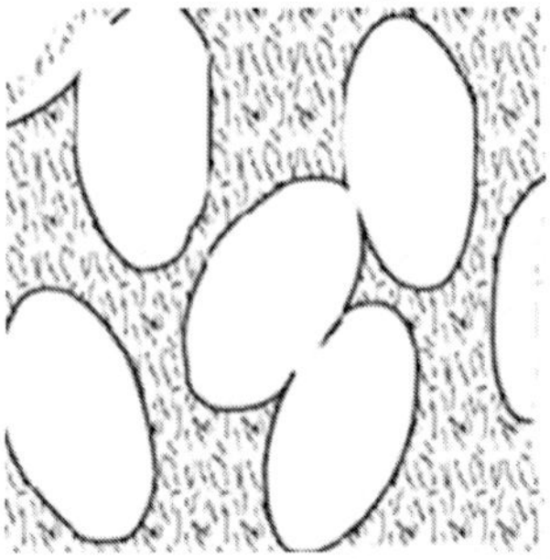
5	Depletion haloes	The porphyroblasts, that are growing during metamorphism may remove the adjacent minerals by reaction process, that show a halo-like appearance surrounding the porphyroblasts 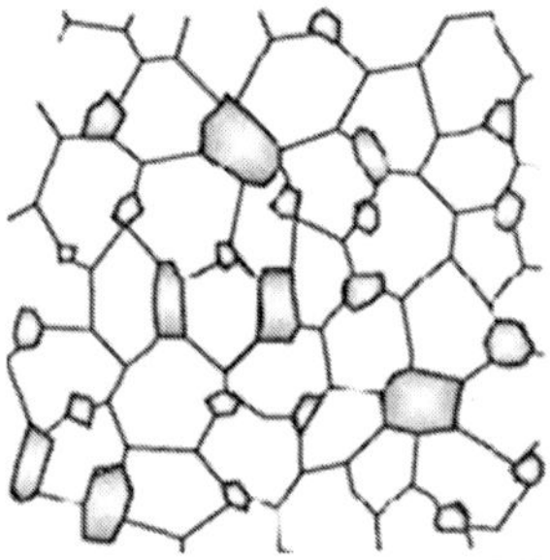
II Non-foliated textures		
6	Granoblastic polygonal	Equidimensional idioblastic to subioblastic crystals of approximately same size and regular polygonal shapes, having fairly straight boundaries meeting at triple junctions, e.g. quartzite having polygonal quartz with mica, calcite and dolomite idioblasts in marble (photomicrograph) 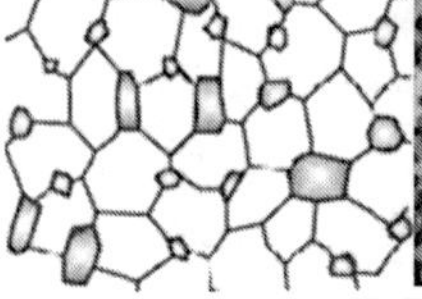 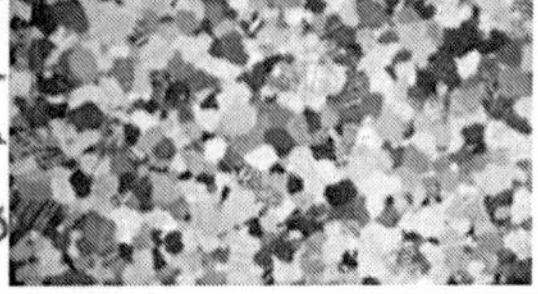

7	Decussate	Randomly oriented interlocking subidioblastic grains, A) pyroxene + amphiboles (B) calcite + dolomite 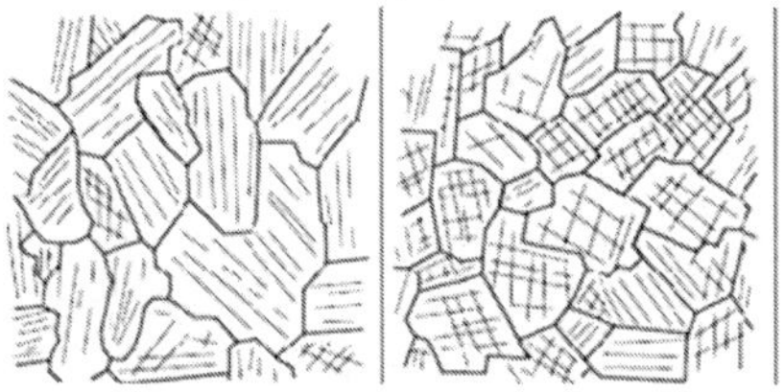 pyroxenes/amphiboles A carbonates B
III Texture of dynamic metamorphism		
1	Cataclastic	The rock that is present in the thrust or fault zone are affected where the minerals are broken into various size fragments and also rotated. The texture is formed by brittle deformation of minerals (it involves grain-size reduction)
2	Mylonitic	It shows foliations alternating with cataclastic and plastic texture. Feldspar and quartz porphyroblasts of granite gneiss are crushed at the margins. Crushed and recrystallised quartz and feldspars are wrapped around porphyroclasts 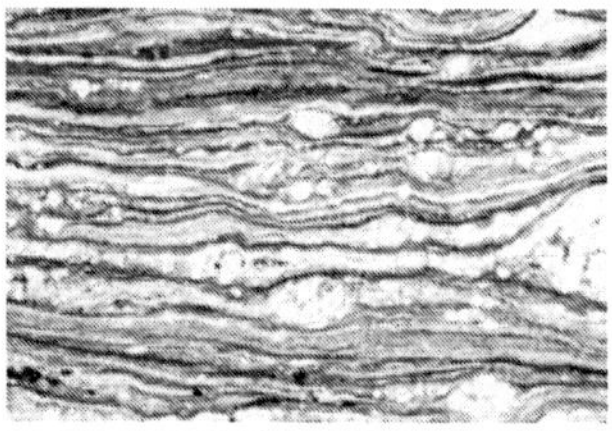
3	pseudotachylite	In advance stage of mylonitization, metamorphic rock contain glass (black material on the left side) generated by localized shear melting, with fine-grained mineral foliation (at the center and right side—greyish colour) (photomicrograph) 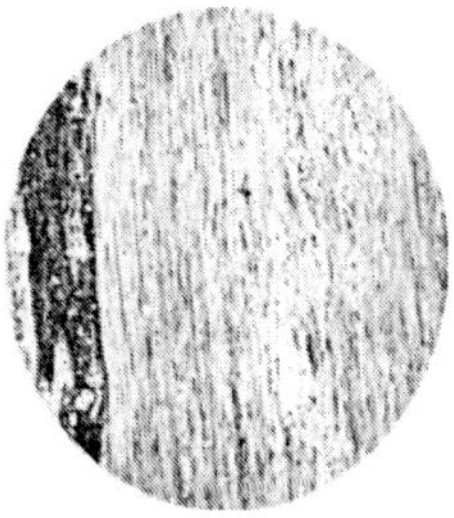

4	Undulose extinction	Some minerals show irregular extinction due to deformation and lattice dislocation in the minerals. Elongate grains of quartz showing undulose extinction due to lattice dislocation (photomicrograph) 
5	Crushed	Quartz crystals are grounded or crushed by shear movement and exhibit fine-grained recrystallised margins derived from quartz prophyroblasts. The proportion of crushed material may vary considerably and even may comprise > 90% of the rock (photomicrograph) 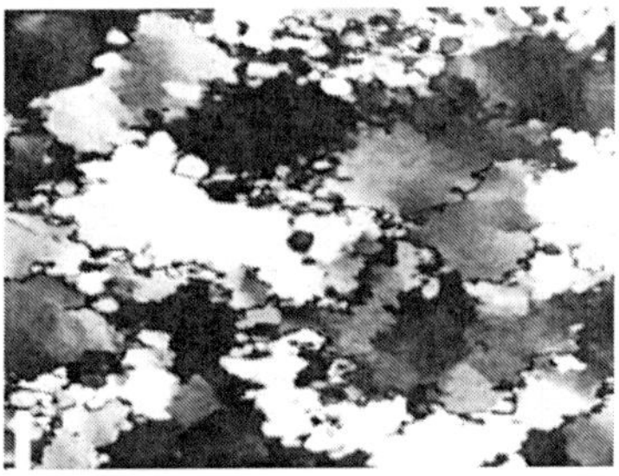
6	Porphyroclast	A large relict crystal in a fine crush matrix
7	Morter	Rounded porphyroblasts are rimmed by crushed matrix
8	Ribbon	Very elongate crystals usually quartz resulting because of cataclasis and intense plastic deformation
9	Kink band	Mica shows kinking due to deformation, as the mineral is malleable. Such bent minerals show undulose extinction and variable interference of 2nd order colour 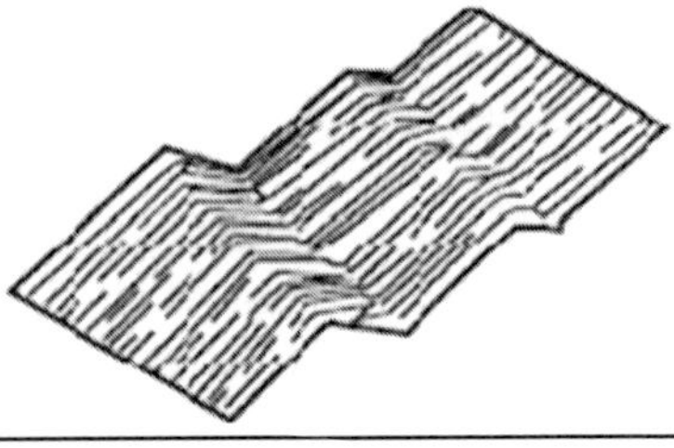

10	Deformation twin	Twinning lamellae in plagioclase feldspars and carbonates are common. When they are deformed, part of the twin lamellae is erased, showing undulose extinction. Drawing shows erased twinning in plagioclase   Photomicrograph of hypersthene granulite showing undulose extinction in feldspars (partially erased twinning lamellae)
11	Oblique foliation	Foliations that cut across the foliations developed due to shear offsets.  Photomicrograph showing angular foliation (greyishwhite colour) merging with east-west (dark grey to blackish) foliation

<table>
<tr><td>12</td><td>Shear band cleavage</td><td>A texture that has shear bands (or C foliation), which are spaced cleavage that transect well-developed mineral foliation. (S foliation) at a small angle
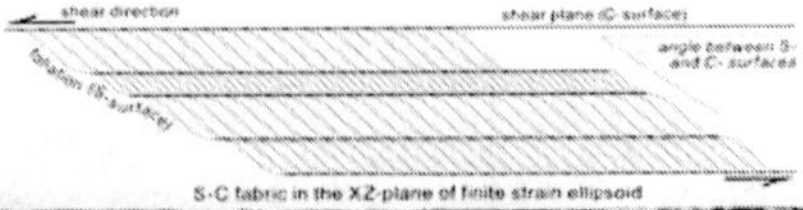

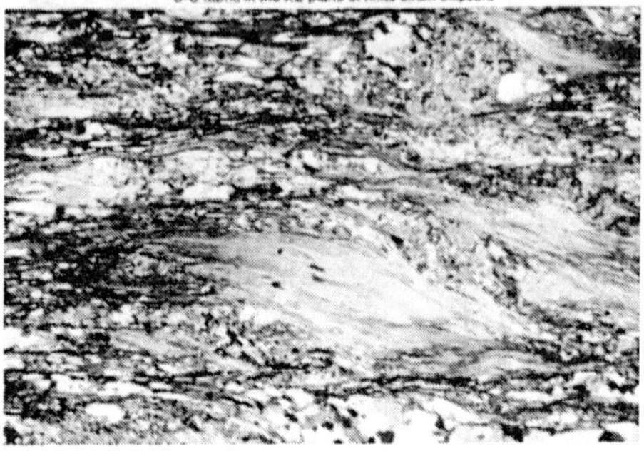
Photomicrograph of mica is showing both S and C fabric. Muscovite (S) second order yellowish colour is bent and merging with (C) shear foliation. As seen in the above drawing</td></tr>
<tr><td>13</td><td>Augen, flaser</td><td>Eye-shaped feldspar porphyroclasts surrounded or wrapped by biotite foliation in augen gneiss (field photo)
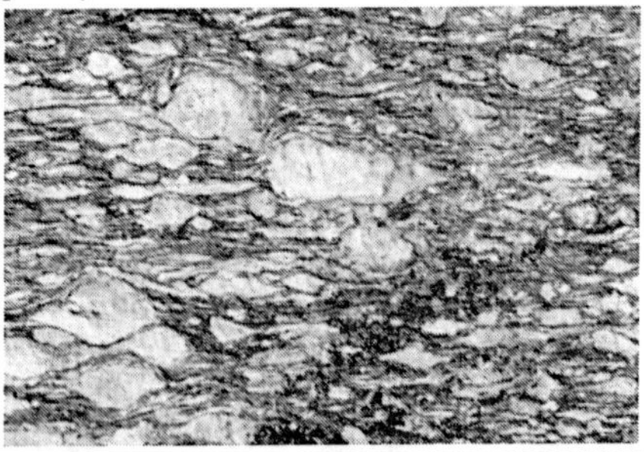</td></tr>
<tr><td colspan="3">IV Texture of regional metamorphism</td></tr>
<tr><td>1</td><td>Spaced foliation</td><td>A foliation developed in zones separated by non-foliated microlithons of non-foliated minerals such as quartz and feldspar</td></tr>
<tr><td>2</td><td>Continuous foliation</td><td>A foliation that is not spaced, but occurs continuously</td></tr>
<tr><td>3</td><td>Cleavage</td><td>Any type of foliation in which the aligned platy phyllosilicates are too fine to recognise with the unaided eye</td></tr>
</table>

4	Stay cleavage	Fine-grained brownish clay rock. In microscope, it shows the cleavage, as in second figure.  Sericite mica and other platy minerals with fine-grained quartz showing slaty cleavage of the clay rock under the microscope 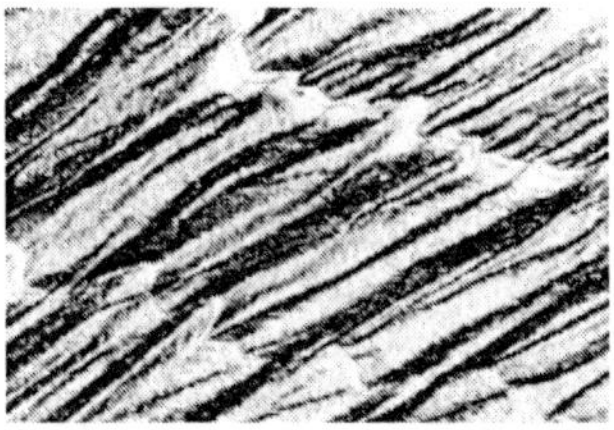
5	Crenulation cleavage	A schistosity that becomes microfolded. Fine grained phyllite show the crenulation cleavage Left photomicrograph–cleavage N-S – almost black color; 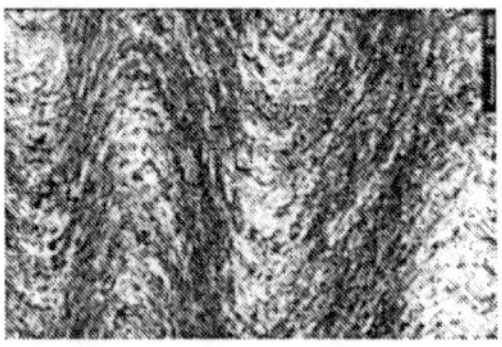 Right photomicrograph – cleavage NNW-SSE – alternating black and yellowish colour 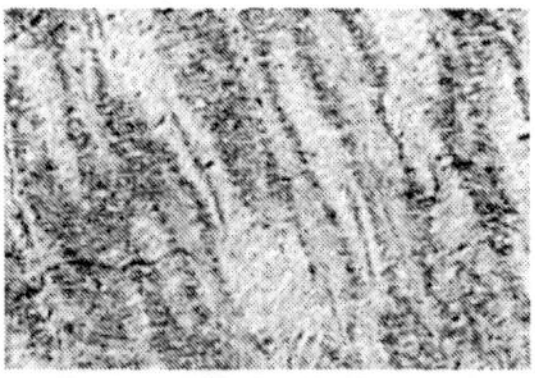 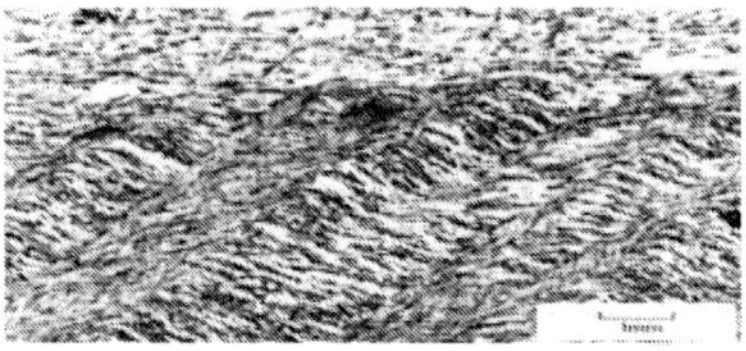 The microfold axes of crenulation cleavage is another form of lineation

6	Schistosity	A planer orientation of elongated mineral aggregates produced by metamorphism. Aligned minerals are coarse enough to see with the unaided eye. Biotite and muscovite minerals show continuous schistosity and folding (photomicrograph) 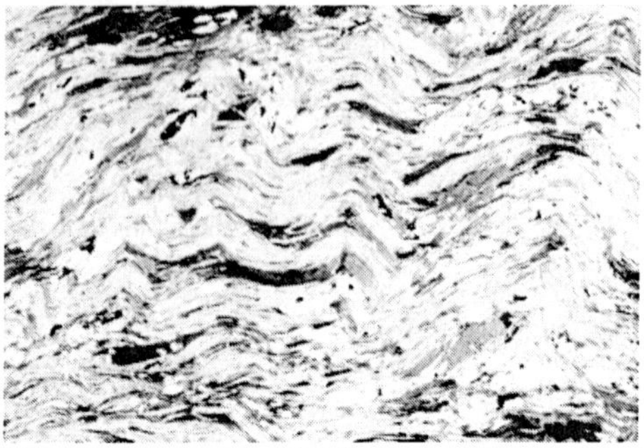
7	Gneissose structure	Layered metamorphic rocks showing poorly developed schistosity. Alternate bands of quartzofeldspathic minerals with thin layers/bands of biotite/hornblende
8	Layering, banding	A foliation consisting of alternating layers of different composition, e.g. mica or amphibole layer is alternating with quartz feldspar layer
9	Prekinematic	Minerals that were present prior to deformation/ metamorphism are deformed. a. Bent quartz with undulose extinction and curved fracture, b. Foliation wrapped around a porphyroblast  c. Pressure shadow or fringe, d. Kink bands or folds in mica, e Microboudinage of quartz, f. Deformation twins of plagioclase

10	Synkinematic	Minerals that grew during deformation/metamorphism are deformed
11	Postkinematic	Minerals that grew after deformation are idioblastic to subidioblastic in nature 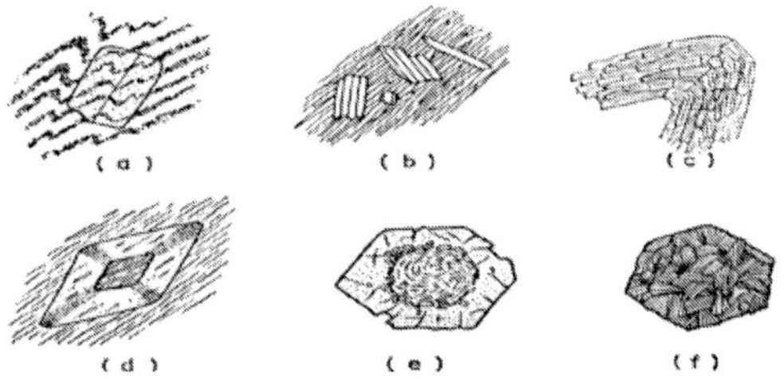 a. Helicitic folds with idioblastic mineral, b. Randomly oriented minerals on the foliated matrix, c. Polygonal arc fold, d. Chiastolite-andalusite having carbonaceous inclusions oriented like + symbol, e. Late inclusion-free rim on poikiloblast(?) f. Random aggregate of pseudomorph mineral
12	Pressure shadow	Concentration of a matrix mineral in low-stress areas adjacent to a porphyroblast (diamond shape) due to pressure solution (solution transfer) and re-precipitation (elongated arrow on both sides of the porphyroblast) 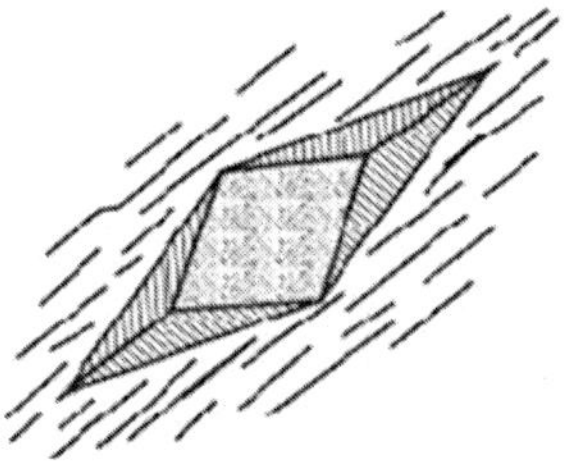
13	Wrapped	A foliation compressing pre-kinematic garnet porphyrobast (or possibly bowed out by the growing porphyroblast) 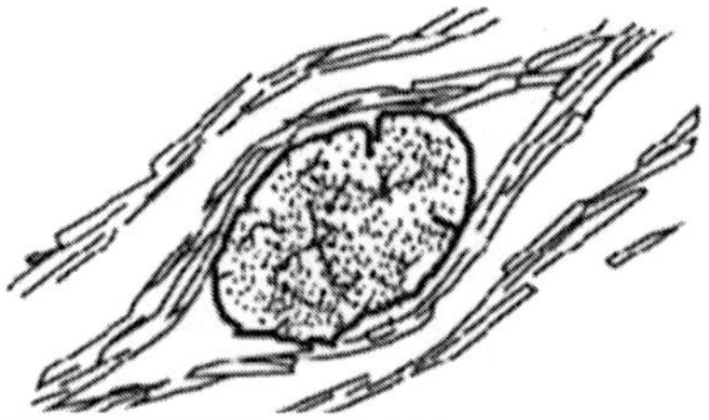

14	Helictic	Folded internal S shape inclusions in garnet porphyroblast cross cutting main foliation indicates a pre-kinematic porphyroblast 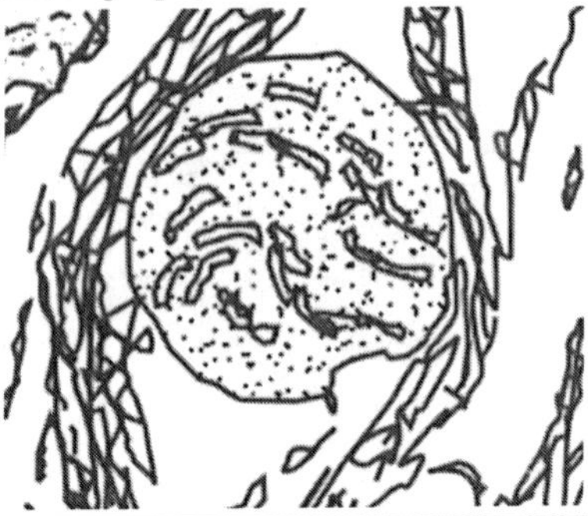
15	Spiral (rotational, snowball)	Spiraled inclusion trails indicate rotation. Foliation from matrix is continuing into garnet porphyroblast as inclusion trails showing S rotation 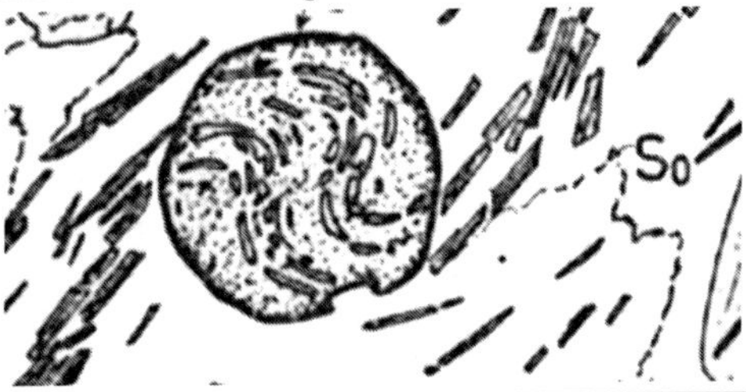
V Other textures		
1	Symplectite	Two or more phases intergrown simultaneously in an irregular manner. These textures are commonly found in high grade metamorphic rocks such as gneiss, granulites, migmatites, etc. Diopside and albite intergrowth (photomicrograph in plane polarized light) 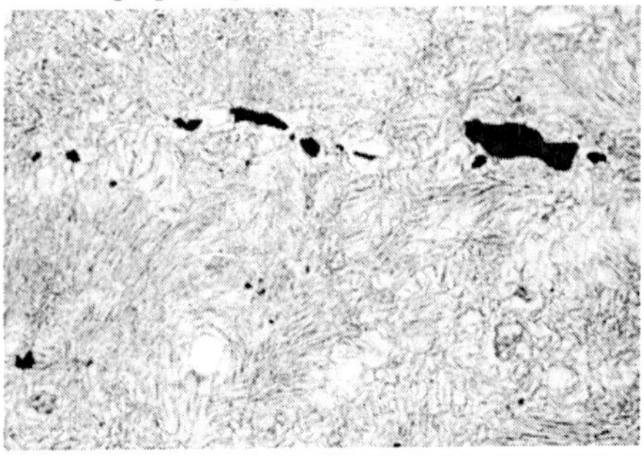

2	Myrmikite	One kind of symplectite containing intergrowth of quartz and feldspar
3	Perthite	Exsolution of plagioclase (dark lamellae) within K-feldspar (White color intertwinned with dark colour)—perthite (photomicrograph) 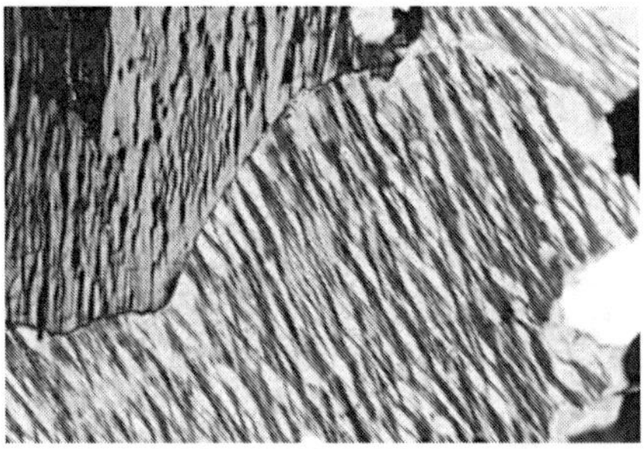
4	Anti-perthite	Exsolution of K-feldspar (dark colored spots) within plagioclase showing twin lamellae (light grey colour) (photomicrograph) 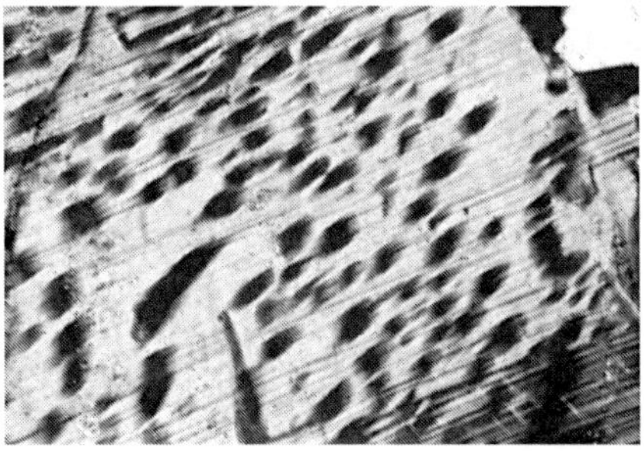
5	Corona	At the contact of sillimanite (fractured white colour) and cordierite (colourless or white), corona growth of sapphirine. (green) as corona (photomicrograph) 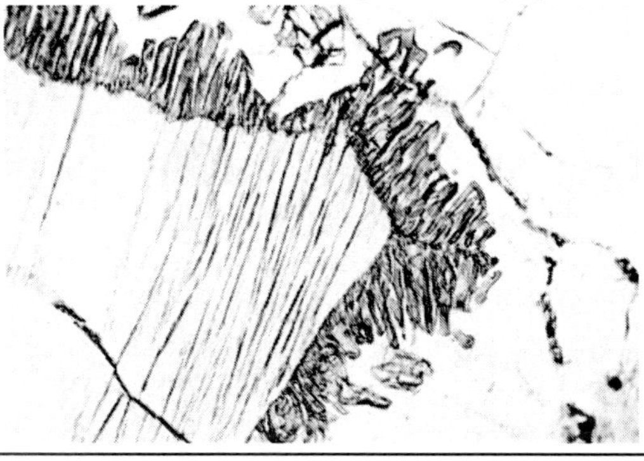

Reference Books

An Introduction to Metamorphic Petrology. ELBS Longman Group, U.K. 242 pp.

Krauskopf, K.B., 1979. *Introduction to geochemistry*. McGraw Hill, Kokakushe Ltd, Tokyo, Japan. 601 pp.

Philpotts, A.R., 1994. *Principles of Igneous and Metamorphic Petrology*. Prentice Hall Pvt. Ltd., Indian edn., New Delhi. 485 pp.

Sharma R.S., 2016. *Metamorphic Petrology—Concepts and Methods*. Geological Society of India, Bengaluru. Textbook series in geological sciences for graduate students. 420 pp.

Spear F.S., 1993. Metamorphic phase equilibria and pressure-temperature-time paths. Monographh I, Mineralogical Society of Ameria, Washington D.C, Chapters 12 and 13.

Spry, A., 1969. *Metamorphic Textures*. Pergamon, Oxford, UK.

Winkler, H.G.F., 1979. *Petrogenesis of Metamorphic Rocks*. Springer-verlong, Heidelberg, Germany, 5th edn. 348 pp.

Winter, J.D., 2012. *Principles of Igneous and Metamorphic Petrology*. Prentice Hall, New Jercy, 2nd Indian edn. by PHI Learning Pvt. Ltd., New Delhi. 693 pp.

Wood, B.J., Fraser, D.G., 1976. *Elementary Thermodynamics for Geologists*. Oxford University Press, London. 299 pp.

Yardley, B.D., 1990. *An Introduction to Metamorphic Petrology*. ELBS Publication, Longman Singapore Publisher (Pvt.) Ltd., Singapore. 248 pp.

Index

T

U

V

W

X

X